D0991791

High Alumina Cement Concrete

Books by the same author:

Properties of Concrete (1972): Pitman (UK); Wiley (USA)

Structural Analysis (1972): International Textbook Company

Hardened Concrete: Physical and Mechanical Aspects (1971): American Concrete Institute

Creep of Concrete: Plain, Reinforced and Prestressed (1970): North Holland

Basic Statistical Methods for Engineers and Scientists (1964): International Textbook Company

High Alumina Cement Concrete

by

Adam Neville

Professor and Head of Department, Department of Civil Engineering, The University of Leeds

President of The Concrete Society

Chairman of the RILEM Permanent Commission on Concrete

in collaboration with

Dr. P.J.Wainwright

Lecturer in Civil Engineering, The University of Leeds

A HALSTED PRESS BOOK

JOHN WILEY & SONS
NEW YORK

To my Cousins-in-law

Library of Congress Cataloging in Publication Data

Neville, Adam M
High alumina cement concrete.

"A Halsted Press book."
Bibliography: p.
Includes index.
1. Concrete. 2. Alumina cement. I. Wainwright, Peter John, joint author. II. Title.
TA439.N47 1975 666.95 75-14379
ISBN 0-470-63280-1

Published in Great Britain in 1975 by:

The Construction Press Ltd.,
Lunesdale House,
Hornby,
Lancaster, LA2 8NB.

Published in the USA in 1975 by:

Halsted Press, a Division
of John Wiley & Sons, Inc.,
New York.

Printed in Great Britain by
The Blackburn Times Press.

Contents

Preface

The idea of this book originated with the publishers who felt that the confusion about the safety of structures made with high-alumina cement concrete has reached a stage when a full and coherent explanation was necessary. Reports about schools closed and roofs propped were rampant and there were as many engineers busy testing existing structures as there were designing new ones in traditional, *not* high-alumina cement, concrete.

This was the aftermath of the failure at the Sir John Cass School in Stepney and, to some extent, of two earlier failures. Although I was involved in the investigation of only one of these structures, perhaps it was thought that, because I had highlighted the inherent dangers of the structural use of high-alumina cement in 1963, I was the man to say 'I told you so'. This I refuse to do, but I agree that builders, architects, engineers, school governors, and indeed the public at large ought to find out more about high-alumina cement concrete structures and about the high-alumina cement problem which is affecting much of this country. Hence the present book, and I hope that it will serve the purpose of placing on record the facts and background of the 'HAC problem'.

I am most grateful to the numerous government departments, local authorities, manufacturers, and individual engineers and architects who provided information and allowed visits to many interesting buildings. The use of published data is acknowledged by the conventional system of references but I would like to express especial thanks to those who allowed me to see internal reports and confidential documents. I am grateful to the Institution of Civil Engineers for permission to draw on my 1963 paper No. 6652 and to the Director of the Building Research Establishment for permission to reproduce plates 6.1, 6.2, 7.1 and 7.2.

The rapid production of this manuscript would not have been possible without the organizing ability of Miss Sue Chapman, who was ably assisted by Miss Christine Morris, Mrs. Anne Haynes, Miss Jane Dransfield and Mrs. Gertrud Uddin. The figures were prepared by Mr. B. Firth, who managed to achieve order and, dare I say, attractiveness out of graphs and sketches from many sources. Messrs. J. J. Brooks and G. A. Hirst kindly proofread the manuscript. Miss A. Buxton, who during the few weeks when the book was being written became Mrs I. Turner, was most assiduous in hunting references, eliminating errors, and generally ensuring the production of the mansucript. To all these people, who worked speedily and efficiently through many hours of 'Monday to Sunday', I am sincerely grateful.

A. M. Neville

Leeds to London commuter train
January, 1975

1 Basic Facts about High-Alumina Cement

1.1 What is HAC?

The name high-alumina cement, for brevity HAC in this book, is well known even to the man in the street as a certain notoriety has been acquired by the material owing to the failures and subsequent investigations in 1973 and 1974. HAC is a cement in that it is used in making concrete but it is also a cementitious material used for other purposes, which are briefly mentioned in Section 2.3.

Portland cement

It is important to realize that HAC is substantially different from what one might call *the* cement: *Portland cement*. Portland cement is by far the most common binder used in the manufacture of concrete, and is a material that has served us well over a great many years. Indeed, 1974 saw the 150th anniversary of the patent for Portland cement, granted to Joseph Aspdin.

While there are many types of Portland cement, they all have one feature in common: they are obtained from silica-, alumina-, and iron oxide-bearing materials burnt at a clinkering temperature (i.e. below the point of complete fusion) so that the main compounds in Portland cement are:

Tricalcium silicate $3CaO.SiO_2$
Dicalcium silicate $2CaO.SiO_2$
Tricalcium aluminate $3CaO.Al_2O_3$
Tetracalcium aluminoferrite $4CaO.Al_2O_3.Fe_2O_3$

High alumina cement

Now, the main compounds in HAC are substantially different. They are calcium aluminates of low basicity: $CaO.Al_2O_3$ and $5CaO.3Al_2O_3$. Thus the lime-alumina (CaO to Al_2O_3) ratio ranges between 1 and 1.67, compared with 3 in the case of tricalcium aluminate in Portland cement, which contains also high lime-silica ratio compounds. A high ratio of lime to silica or to alumina leads to alkalinity, and this is of importance with respect to the alkali resistance of the material and to the protection of the steel reinforcement from corrosion. These matters are considered in Section 5.4.

The quantities of the compounds present in cement are generally not determined direct but are calculated from the known proportions of the oxides. While this is relatively straightforward in Portland cement, in the case of HAC no simple method of calculation is available, and indeed considerably less is known about the compound composition of HAC than of Portland cement. Table 1.1 gives the range of typical percentages of the oxides present in HAC. A minimum alumina content of 32 per cent is prescribed by the British Standard for High Alumina Cement BS 915: Part 2: 1972, which requires also the alumina-lime ratio to be between 0.85 and 1.3.

Table 1.1 **Range of typical percentages of oxides present in HAC**

Oxide	*Content (per cent)*
CaO	35 to 39
Al_2O_3	37 to 41
SiO_2	3½ to 5½
Fe_2O_3	9 to 12
FeO	4 to 6
TiO_2	1½ to 2½
MgO	½ to 1
Insoluble residue	1

As already mentioned, the exact nature of the compounds actually present in HAC has not been well established, but there is no doubt that the most important compound present is monocalcium aluminate $CaO.Al_2O_3$. The other calcium aluminate present was referred to earlier as $5CaO.3Al_2O_3$. This is the traditional formula although the actual composition is in all likelihood $12CaO.7Al_2O_3$ (which contains also FeO, SiO_2 and TiO_2 [1]). With the latter composition, the lime-alumina ratio would be 1.71 rather than 1.67. For our purposes, this is not particularly significant, and it is only fair to admit that the simple stoichiometric compounds of elementary chemistry are not necessarily to be expected in the case of the complex compounds found in cement.

Details of the chemistry of HAC have to be sought elsewhere (see, for instance, Refs. 2 and 3). Here it suffices to say that in addition to $CaO.Al_2O_3$ and $12CaO.7Al_2O_3$, HAC contains some $CaO.2Al_2O_3$, $2CaO.SiO_2$, $2CaO.Al_2O_3.SiO_2$, and $CaO.TiO_2$.

1.2 Hydration of HAC

HAC, like Portland, is a hydraulic cement, i.e. it sets and hardens by chemical reactions with water and remains stable in the presence of water. Let us briefly look at some of the reactions involved.

Reactions

The hydration of $CaO.Al_2O_3$, which, of the compounds present in HAC, has the highest rate of strength development, results in the formation of $CaO.Al_2O_3.10H_2O$, a small quantity of $2CaO.Al_2O_3.8H_2O$, and alumina gel (Al_2O_3.aq). The hexagonal* hydrate $CaO.Al_2O_3.10H_2O$ is of interest in connexion with conversion (about which much will be said in Chapter 3) because this hydrate is unstable both at normal and at higher temperatures and becomes transformed into cubic crystals $3CaO.Al_2O_3.6H_2O$ and alumina gel. The formation of the cubic* hydrate is encouraged by a higher concentration of lime or a rise in alkalinity [4].

*These terms refer to the symmetry of the crystal system and not to the shape of the products of hydration.

The compound $5CaO\ 3Al_2O_3$ or $12\ CaO.7Al_2O_3$ is believed to hydrate to $2CaO.Al_2O_3.8H_2O$. The product of hydration of $2CaO.SiO_2$ is a calcium silicate hydrate, in which the number of molecules of water is not known with certainty. The reactions of the hydration of the other compounds, particularly those containing iron, have not been determined with any degree of certainty, but the iron held in the glass (non-crystallised part of cement) is known to be inert[5].

Water of hydration

The water of hydration of HAC, i.e. the amount of water which combines chemically with the anhydrous cement during hydration, is about 50 per cent of the weight of the cement[2]. This is approximately twice as much as the water required for the hydration of Portland cement. There is no fundamental significance in this insofar as the behaviour of concrete is concerned, but the high water of hydration of HAC was the reason for the high water-cement ratios (not less than 0.5) specified in the past. This recommendation was abandoned some fifteen years ago.

1.3 Historical note

HAC was developed as a solution to the problem of decomposition of Portland cement under sulphate attack. The inventor was a Frenchman, Jules Bied, who realized that calcium aluminates would provide the necessary resistance which is absent in calcium silicates. The patent was taken out in France in May 1908 and was registered in the United Kingdom (under the number 8193) a year later.

Difficulties in quality control delayed the production of HAC until 1913, but the cement did not enter the general market (because of further trials) until the end of the First World War. In the meantime, it was realized that HAC develops a very high early strength, which commended it for use in the construction of gun emplacements during the First World War, and then to other structural uses. Subsequently, the manufacture of the cement extended to the United States in 1924 and to the United Kingdom in 1925. Here, the first works was established by the British Portland Cement Manufacturers Ltd. in Magheramorne in Northern Ireland, using local bauxite and the rotary kiln process. The production of HAC there ceased after a short period of operation. In 1926, a new works was opened in West Thurrock in Essex by the Lafarge Aluminous Cement Company Ltd., and this is the only British works.

Later on, the manufacture of HAC extended to other countries, notably Germany, Hungary, Czechoslovakia, Italy (the works passing to Yugoslavia after the Second World War), Japan, Spain, and USSR. HAC is, of course, used also in many other countries.

While Portland cement is described throughout the world by the name Portland, HAC enjoys a wide variety of names, many of them trade names. Indeed, even in the United Kingdom, there are two basic names: high-alumina cement and aluminous cement. There are also two trade names, depending on the marketing organization (but all cement comes from the same manufacturer): *Ciment Fondu* and *Lightning* cement. The name Ciment Fondu owes its existence to the fact that, unlike Portland cement, HAC is obtained by solidification of a completely fused mass in the process of manufacture.

Figure 1.1 **Diagrammatic representation of the manufacture of HAC.**

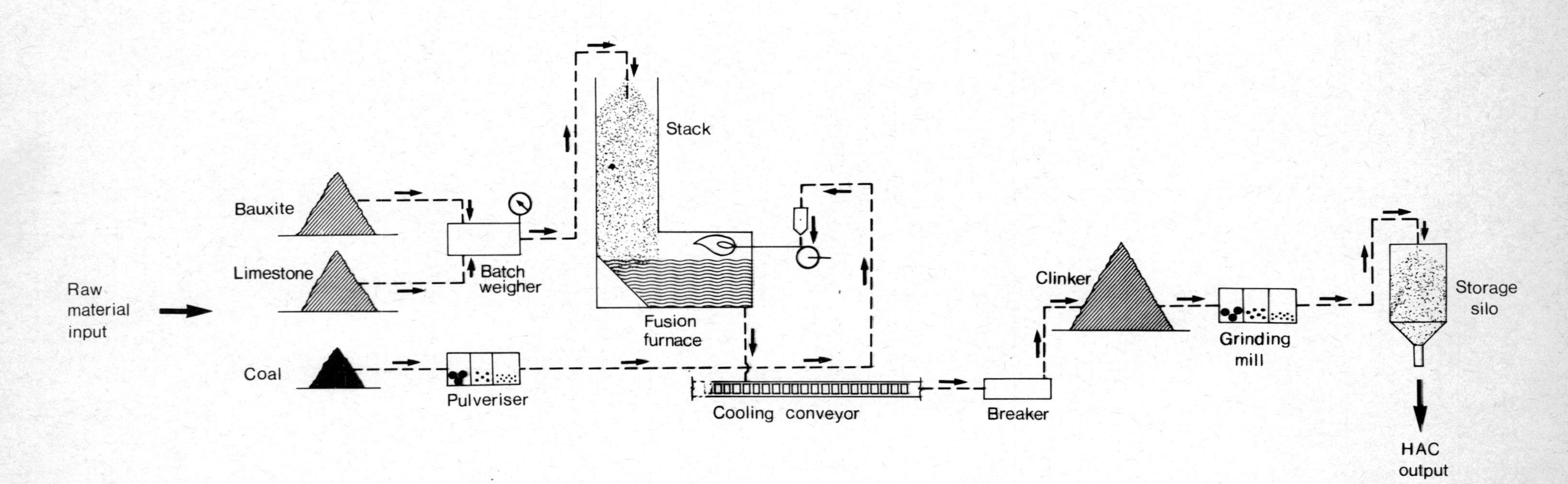

Some of the foreign names may be of interest: *ciment alumineux* in France and *Tonerdezement* or *Tonerdeschmelzzement* in Germany. Many of the other names are a translation of one or the other of these. In the United States, the trade name is *Lumnite.*

Clearly, the cements manufactured in different countries differ in their raw materials, method of manufacture, and, to some extent, in their properties. These are discussed in Section 2.1.

1.4 Manufacture of HAC

Raw materials

There are two major differences between the processes of manufacture of HAC and of Portland cement. For the latter we use as raw materials normally limestone or chalk and an argillaceous material, usually clay or shale. The raw materials used in the manufacture of HAC are limestone (or chalk) and bauxite, which is a naturally occurring material (a product of weathering of igneous rocks under tropical conditions), very rich in alumina.

Heat treatment

The second difference concerns the actual heat treatment. In the manufacture of Portland cement, the raw materials, suitably comminuted, are heated to a point of near fusion, known as clinkering, but they do not actually become molten. On the other hand, in the case of HAC the material in the furnace becomes completely fused, and indeed this is the origin of the trade name Ciment Fondu in France and in the United Kingdom.

The preceding description applies to the British method of manufacture of HAC, also used in some other countries. Somewhat different processes are used elsewhere, depending on the exact nature of the raw materials.

Manufacturing process

A diagrammatic representation of the British process of manufacture of HAC is shown in Fig. 1.1. The reverberatory furnace is L-shaped, this being a development of the original French *water-jacket* process. The raw materials (bauxite and limestone), are fed in predetermined proportions, about 2 tonnes at a time, into the vertical stack of an open-hearth furnace in lump form (up to 100 mm) in order to allow a reasonably free passage of the combustion gases. This is also assisted by means of exhaust fans. As it is important that the materials be fed in lump form, any bauxite fines in the processed raw material have to be mixed with small quantities of cement and water and made into briquettes before they can be charged into the furnace.

The lower part of the stack of raw materials in the furnace is heated by means of hot air from a pulverised coal flame. The weight of the coal used is 22 to 25 per cent of the weight of cement produced. Oil may also be used. In passing down the stack, the raw materials are progressively dehydrated and decarbonated, and the resultant calcium and aluminium oxides eventually melt at about 1600°C and react with one another at the base of the stack.

From here, the molten pool of material flows continuously into the horizontal part of the furnace, from where the molten mass is tapped into cooling conveyors made of heavy steel plate. The molten material, cooled at a controlled rate, forms flat slabs (pigs) which are then broken up to form clinker. The clinker is then taken to stock piles where it is cooled further before being taken to the grinding mills. These mills are similar to the ball mills used in the manufacture of Portland cement.

Table 1.2 **Typical analyses of raw materials used in the manufacture of HAC[3]**

Source of bauxite	Percentage content							
	Combined water	Loss on ignition	SiO_2	Al_2O_3	Fe_2O_3	CaO	TiO_2	MgO
France	5 - 10	11½ - 12½	3½ - 6	45 - 55	20 - 25	½ - 3	2½ - 3½	–
Greece	1 - 3	11 - 12½	2 - 3½	50 - 58	25 - 30	½ - 3	2½ - 3½	–
Yugoslavia	3 - 8	18 - 20	2 - 3	50 - 55	20 - 25	½ - 3	2½ - 3½	–
Mixed (including Surinam)	3½	24	3½	56	14	–	2½	–
Limestone	$\frac{1}{10}$ - 2	42 - 43½	½ - 1½	¼ - 1	$\frac{1}{5} - \frac{3}{5}$	53 - 55½	–	¼ - 1

Table 1.3 **Typical oxide composition of HAC made in various countries[6]**

Country of manufacture	CaO	Al_2O_3	SiO_2	Fe_2O_3	FeO	TiO_2	Colour	Remarks
United Kingdom France Spain Yugoslavia	35 - 39	37 - 41	3½ - 5½	9	6	1½ - 2	Very dark grey	No additions during grinding
Czechoslovakia	35 - 39	39 - 44	3½ - 5½	9 - 14	0 - 1	1½ - 2	Yellow-brown	Sintered, not fused
Germany	38 - 42	44 - 51	5 - 8	0 - 1	½ - 2	1 - 2	Very light grey	Some content of SO_3 and metallic iron
United States	35 - 39	37 - 41	8 - 10	4 - 6	4 - 6	1½ - 2	Light grey	Some blastfurnace slag added during grinding

No material is added to the HAC clinker, unlike the case of Portland cement where interground gypsum is necessary for the retardation of setting time. The setting time of HAC is controlled by its composition.

Due to the high degree of hardness of the HAC clinker, the wear on the screens and mills is heavy and the power consumption is high. This, together with the cost of bauxite, accounts for the relatively high cost of HAC. The bauxite used in England is usually imported from Greece and less frequently from France. Typical analyses of various bauxites are shown in Table 1.2.

During the manufacture of the clinker, the chemical composition of the raw materials and of the end product is continually monitored, and adjustments to the proportions of the raw materials fed into the stack are made as necessary.

As mentioned previously, the manufacturing process differs somewhat in many other countries, depending on the nature of the raw materials used. If the materials are too soft they tend to be crushed under their own weight in the stack, which would prevent the escape of the combustion gases. In such a case, the process employing a rotary kiln, similar to that used in the manufacture of Portland cement, is used. The raw materials are ground together to a fine powder before being fed into the kiln. The kiln is fired by pulverised coal, and the resulting molten material collects at the lower end of the kiln where it is tapped and cooled, and then crushed to form clinker. This process is used in the United States and was also used in the Northern Ireland works.

During the Second World War, shortage of bauxite in Great Britain led to the use of a synthetic source of alumina as a partial replacement in the manufacture of HAC. This substitute was made from waste products in the manufacture of alumina: aluminium dross and red mud.

There exist other manufacturing processes in use in various countries but their details are of no direct relevance. Table 1.3 shows typical chemical analyses of the HACs produced.

Annual production

The annual production of HAC in the United Kingdom is of the order of 120,000 tonnes. This may be compared with some 17 million tonnes of Portland cement manufactured per annum. Of the 120,000 tonnes, about one-third is exported for refractory purposes (see Section 2.3). The British production represents approximately one-fifth of the world production of HAC excluding USSR.

2. Properties and Uses of High- Alumina Cement

2.1 Physical properties

A review of the main physical properties of HAC and a general comparison with Portland cement may be of interest.

Density

The loose bulk density of HAC is not significantly different from that of Portland cement. The specific gravity is also similar, possibly slightly higher, 3.20 to 3.25.

Fineness

Fineness of the cement is specified by the British Standard BS 915: Part 2: 1972 as not less than 225 m^2/kg (2250 cm^2/g) or alternatively the residue on a BS No. 170 sieve (90μm aperture) is to be not more than 8 per cent by weight. The actual fineness is clearly governed by grinding during the manufacture of the cement and is rather higher than the minimum; 300 m^2/kg (3000 cm^2/g) is probably a typical value. This is a somewhat lower fineness than that of modern Portland cements.

Colour

The colour of HAC is grey-black, and thus much darker than that of Portland cement. The colour depends on the amount of iron present and on its state of oxidation: the more iron there is, and the more in ferrous form, the darker the colour. Ferric compounds lead to a brown colour [3]. The colour may be of architectural interest but does not affect the strength properties of HAC concrete.

We may note that there exists also a white HAC used for castable refractories. This is made using alumina as a raw material and contains 70 to 80 per cent of Al_2O_3 with only very little silica and iron oxide. The white cement has a higher melting temperature than ordinary HAC. White HAC, sold under the name of *Secar,* is very expensive (about £145 per tonne) and its use is limited to special refractory applications.

Compressive strength

The compressive strength of HAC is specified in the British Standard BS 915: Part 2: 1972 by a 1:3 mortar test, using a water-cement ratio of 0.40. The strength is to be not less than 42 N/mm^2 at 24 hours and 49 N/mm^2 at 3 days with the proviso that the 3-day strength must be higher than the 24-hour strength.

Setting time

The setting time is specified in the same standard as between 2 and 6 hours for the initial set; the final set is to occur not more than two hours after the initial set. Typical actual values are 2h 30m for the initial set and 3h 20m for the final set. However, once the reactions of hydration start to take place in

HAC concrete, it begins to stiffen so that the time during which HAC concrete can be worked is somewhat shorter than when Portland cement is used[3].

From the figures given above we can see that HAC is rather slow *setting* but the subsequent development of strength is extremely rapid; HAC is thus ultra rapid *hardening*. Fig. 2.1 shows typical strength-time curves for HAC and Portland cement concretes, each with a water-cement ratio of 0.4.

Figure 2.1

Typical strength development of HAC and Portland cement concrete made with a water-cement ratio of 0.4.

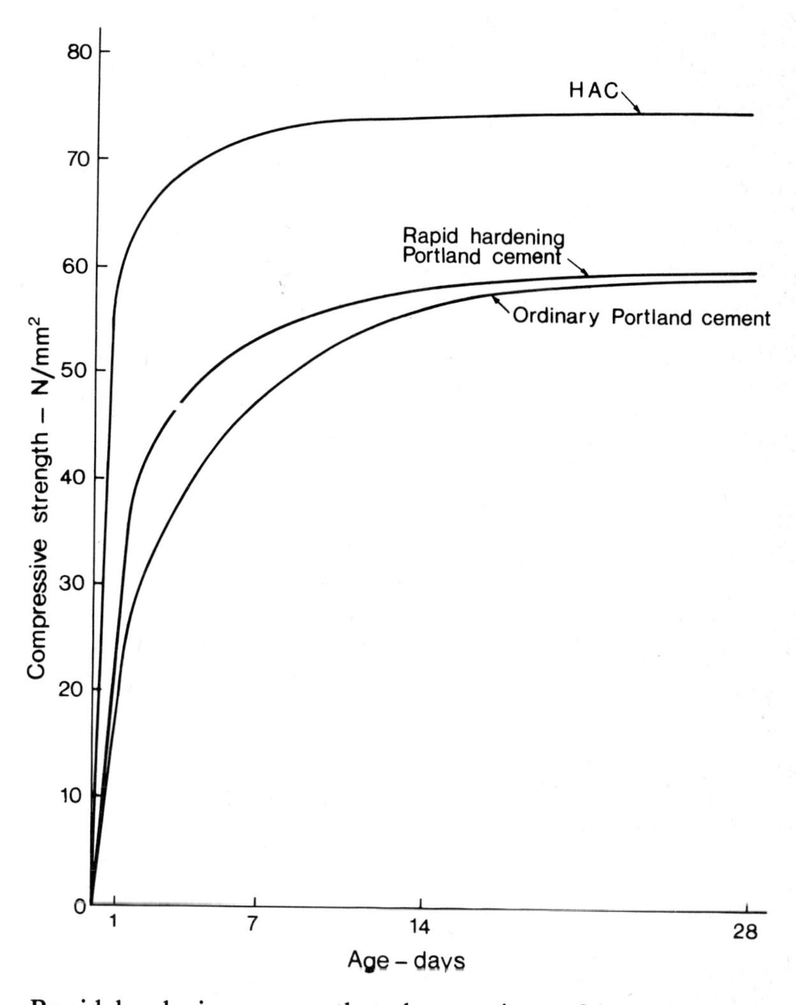

Heat of hydration

Rapid hardening means that the reactions of hydration take place fast. Consequently, the heat evolved in these reactions, known as heat of hydration, is released at a high rate. Table 2.1 gives comparative values of the heat of hydration for HAC and Portland cements[3]. In the case of low-silica HAC, the maximum rate of heat production (shown in the same table) occurs about six hours after mixing so that the maximum temperature in the concrete may occur as early as 8 to 10 hours[3]. The actual value under practical conditions depends of course on the size of the concrete mass.

When no heat loss is permitted, i.e. under adiabatic conditions of curing (which would be virtually the case inside a large mass of concrete), the temperature rise of a 1:2:4 mix with a water-cement ratio of 0.60 was found to be approximately 35 °C at 10 hours, 42 °C at 24 hours and 45 °C at 3 days[3]. The influence of mix proportions on the temperature rise is shown in Table 2.2.

Table 2.1

Comparison of the heats of hydration of different cements[3]

Cement	*Heat of hydration (cal per gram) at*				
	1 day	*3 days*	*7 days*	*28 days*	*90 days*
HAC	77 - 93	78 - 94	78 - 95		
Rapid-hardening Portland	35 - 71	45 - 89	51 - 91	70 - 100	
Ordinary Portland	23 - 46	42 - 65	47 - 75	66 - 94	80 - 105

Table 2.2

Temperature rise of HAC concrete under adiabitic conditions[3]

Cement-aggregate ratio	1:5¼	1:6	1:9	1:12
Temperature rise at 3 days, ^{o}C	50	45	33	23
Total heat released, cal/gram of cement	76	77	82	76

Soundness

HAC contains no free lime so that there is no risk of unsoundness, i.e. delayed appreciable expansion in service. Nevertheless, the British Standard BS 915: Part 2: 1972 prescribes the Le Chatelier soundness test; this is probably of no value.

Porosity

In Section 1.2 we mentioned the large amount of water used up in the reactions of hydration of HAC. One effect of this is that, for the same mix proportions, HAC results in a concrete with a lower porosity and therefore higher impermeability than would be the case with Portland cement. A further observation may be made concerning the water requirement from the workability standpoint. The surface of HAC grains is of lower rugosity, i.e. it

Workability

is smoother, than that of Portland cement. As a result, the same amount of water leads to a higher workability or, conversely, the same workability can be achieved with a somewhat lower water content, other things being equal. However, HAC concrete mixes are somewhat harsher than mixes made with Portland cement, so that with HAC a slightly higher fines content should be used[2].

Shrinkage

Shrinkage of HAC concrete is approximately of the same magnitude as when Portland cement is used but occurs somewhat more rapidly.

Creep

In general, creep of HAC concrete is of similar magnitude to that of similar Portland cement concrete. However, since creep is proportional to the stress-strength ratio, the loss of strength on conversion leads to an increase in creep. Details of creep of HAC concrete are outside the scope of this book but one aspect of creep is worth mentioning. In 1939, it was found[7] that the creep of HAC concrete when stored dry was less than when stored wet; this is contrary to the behaviour of Portland cement concretes, and probably for this reason has been quoted again and again. For instance, there is a mention of this in a book on HAC and HAC concretes published in 1962. And yet, in 1960 we offered an explanation in terms of a possible conversion of HAC[8] (see Chapter 3).

This refers to the fact that when the specimens were subjected to a sustained load for the determination of creep, the wet one had a lower strength than the dry one. The applied stress was the same in both cases so that, allowing for the stress-strength ratio[73], there was very little difference between the creep strains of dry and wet specimens (see Fig. 2.2). No reason for the difference in strength is given in the original paper but it is possible that, inadvertently, conversion occured under humid conditions but not in the air-stored concrete.

Figure 2.2

Variation in the product of creep and strength with time for air-stored and water-stored HAC concrete.

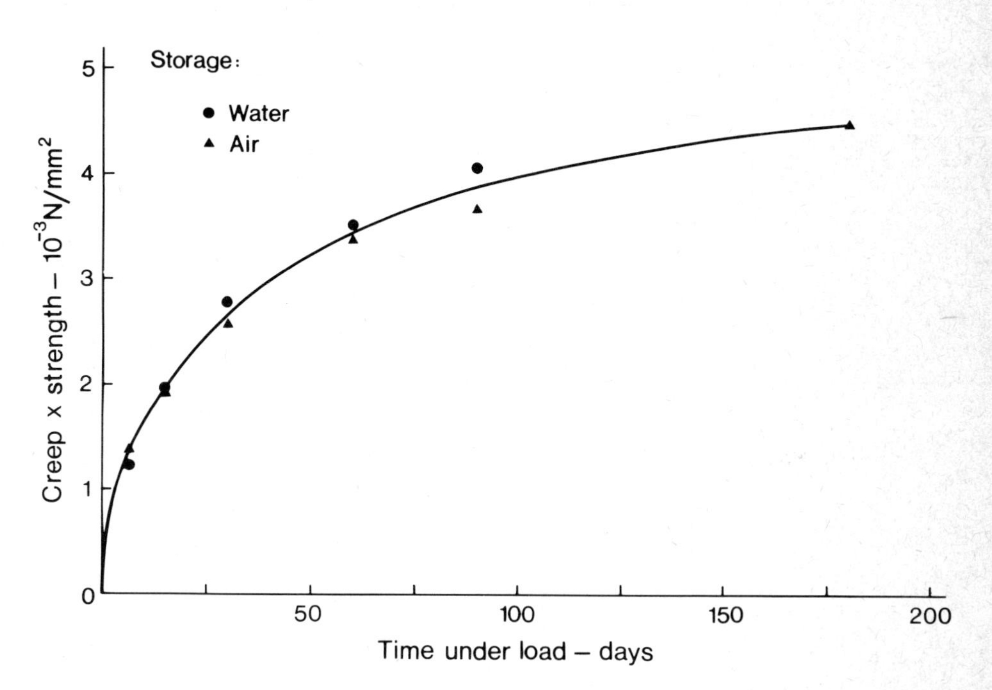

The permeability of HAC concrete, its coefficient of thermal expansion, and modulus of elasticity are all sensibly the same as for concretes of similar mix proportions made with Portland cement[2].

Wet curing of HAC concrete during the first 24 hours is strongly recommended. This provides an efficient means of cooling, but wet curing as such is not essential. This view is contrary to that held previously, which was based on the fact that HAC combines with up to 50 per cent of water on full hydration, and loss of water would decrease the extent of hydration. However, once hardening has occurred, storage in water does not appreciably increase the amount of combined water [24] and for this reason loss of water should be prevented during the first 24 hours after placing. Full hydration is by no means necessary for high strength, provided a dense gel of low porosity is present but a surface allowed to dry at an early stage is prone to dusting.

That it is not the absence of active wet curing that causes dusting, however, is shown by the fact that the surfaces of a specimen in contact with the mould do not dust, and only the unformed top surface does so. Thus, dusting is probably due to carbonation [24], which is facilitated by rapid drying.

Curing

Wet curing is nevertheless the most effective means of reducing the rise in the temperature of the concrete owing to early hydration. This is illustrated by the data of Table 2.3. Here, two groups of 100mm by 240mm cylinders made at 10°C of concrete with a water-cement ratio of 0.48, were cured during the first 24 hours in water at 5°C and in moving air at the same temperature, respectively. From the age of 24 hours onwards all specimens

Table 2.3 **Influence of curing medium on temperature and strength of HAC concrete**[19]

Storage condition during first 24 hours		*In water at 5°C*	*In air at 5°C*
Temperature at the age of	0 h	10	10
	3 h	-	6
	4½h	9	21
	6½h	17	48
	24 h	8	6
Age at test, *days*		7	28
Compressive strength (N/mm^2)		60	20

were stored in air at 5°C. The maximum temperature within the concrete was 48°C for the air-cured concrete and only 17°C for the water-cured specimens; these values indicate the efficiency of water-curing as a means of cooling. The loss of strength due to the rise in the temperature in the air-cured specimens was high (see Table 2.3 and Chapter 3).

2.2 Chemical resistance

The calcium aluminates in HAC, unlike the calcium silicates in Portland cement, do not liberate free calcium hydroxide during hydration and hardening. Since it is this constituent in hydrated cement paste that is attacked by a variety of substances, its absence is the reason why HAC is resistant to many forms of chemical attack. The aluminium hydroxide gel (Al_2O_3.aq), which is produced during hydration, is, however, attacked by caustic alkalis such as sodium and potassium hydroxide, and HAC is, therefore, not resistant to such agents.

As mentioned in Section 1.3, HAC was developed to be a sulphate-resisting cement. In addition to its good performance in suphate-bearing ground and in sea water, HAC is able also to resist many forms of dilute acid attack. Examples are given below.

Type of acid	*Industrial use*	*Concentration*
Sulphuric	gas-washing plant	above pH4
Sulphurous	bleaching chambers	
Lactic	dairy	up to 1 per cent
Tannic	tannery	
Humic	peaty waters	
Carbonic	mineral waters	
Butyric	brewery	

Other substances to which HAC shows great resistance include sugar, fruit juices, fish liquors, animal wastes and blood. Reasonable resistance is also offered to: ground contaminated by chemical refuse dumps (provided there are no concentrated acids present), cyanides, bleaching chemicals (including chlorine bleach), carbon disulphide, sodium bisulphate, photographic chemicals (provided no caustic alkalis are present), glycols, glycerine, phenols, molasses, vinegar (but not strong acetic acid), oils, animal fats, and many others[10].

We should note that lean mixes (aggregate-cement ratio of 9 or more) do not show the characteristic resistance to chemical attack[2].

The resistance of HAC to sulphate attack is decreased by conversion (see Chapter 3), especially in the case of attack by magnesium sulphate[2]. Tests have shown that the permeability is increased by conversion and so is the resistance to freezing and thawing[131]. The reason for this is probably the increase in porosity on conversion. Resistance to acids seems to be unimpaired by conversion[2], probably because no chemical reactions take place.

Unlike Portland cement, HAC does not attack aluminium or lead so that inserts of those metals can be used in HAC concrete.

2.3 Refractory use

At temperatures in excess of about 300°C the strength of concrete made with Portland cement begins to be significantly reduced. Between 400 °C and 450 °C,calcium hydroxide becomes dehydrated, and beyond 600 °C there is a very serious reduction in strength, even with the best aggregate. The most fire-resistant concrete can withstand exposure to 900°C only for a short period[2].

High temperature behaviour

The behaviour of HAC concrete differs in some important respects. Its performance above room temperature and up to about 500 °C is inferior to that of Portland cement concrete, then up to 800 °C the two are comparable, but at above about 1000 °C HAC gives excellent performance. Fig. 2.3 shows the behaviour of HAC concrete made with four different aggregates over a temperature range up to 1100 °C[11]. It can be seen that concretes made with expanded shale and anorthosite aggregates lost about half of their strength after exposure to 400°C. The compressive strengths reached a minimum of 26 and 18 per cent of the original, respectively, after firing at 1000°C. A slight gain in strength occurred for the anorthosite concrete at 1100°C. The compressive strength of phonolite concrete specimens reached a minimum of 8 per cent at 400 °C. Between 400 °C and 1000 °C their strength recovered, reaching a value of 35 per cent at 1000°C. Ilmenite concrete lost more than 60 per cent of its strength after exposure to 100°C. After firing at 1000°C, the residual strength of ilmenite concrete was only 4.5 per cent[11]. The gain in strength at 1000 °C and 1100 °C for phonolite and anorthosite concretes, respectively, is of interest as it is due to the development of ceramic bond.

Ceramic bond

This requires some explanation. When mixed with water, HAC hardens by the reactions of hydration (i.e. with water) and develops what is known as hydraulic bond. This is similar to what happens with Portland cement. The bond is reduced at higher temperatures (see Fig. 2.3) and a compressive strength as low as 7 N/mm^2 may be reached even by a high quality HAC concrete. However, from about 700°C upwards, depending on the type of aggregate, a new form of bond, known as a ceramic bond, begins to form. The bond is established by solid reactions between the cement and the fine aggregate, and increases with an increase in temperature and with the progress of the reactions.

Figure 2.3 **Strength of HAC concretes made with different aggregates as a function of temperature[11].**

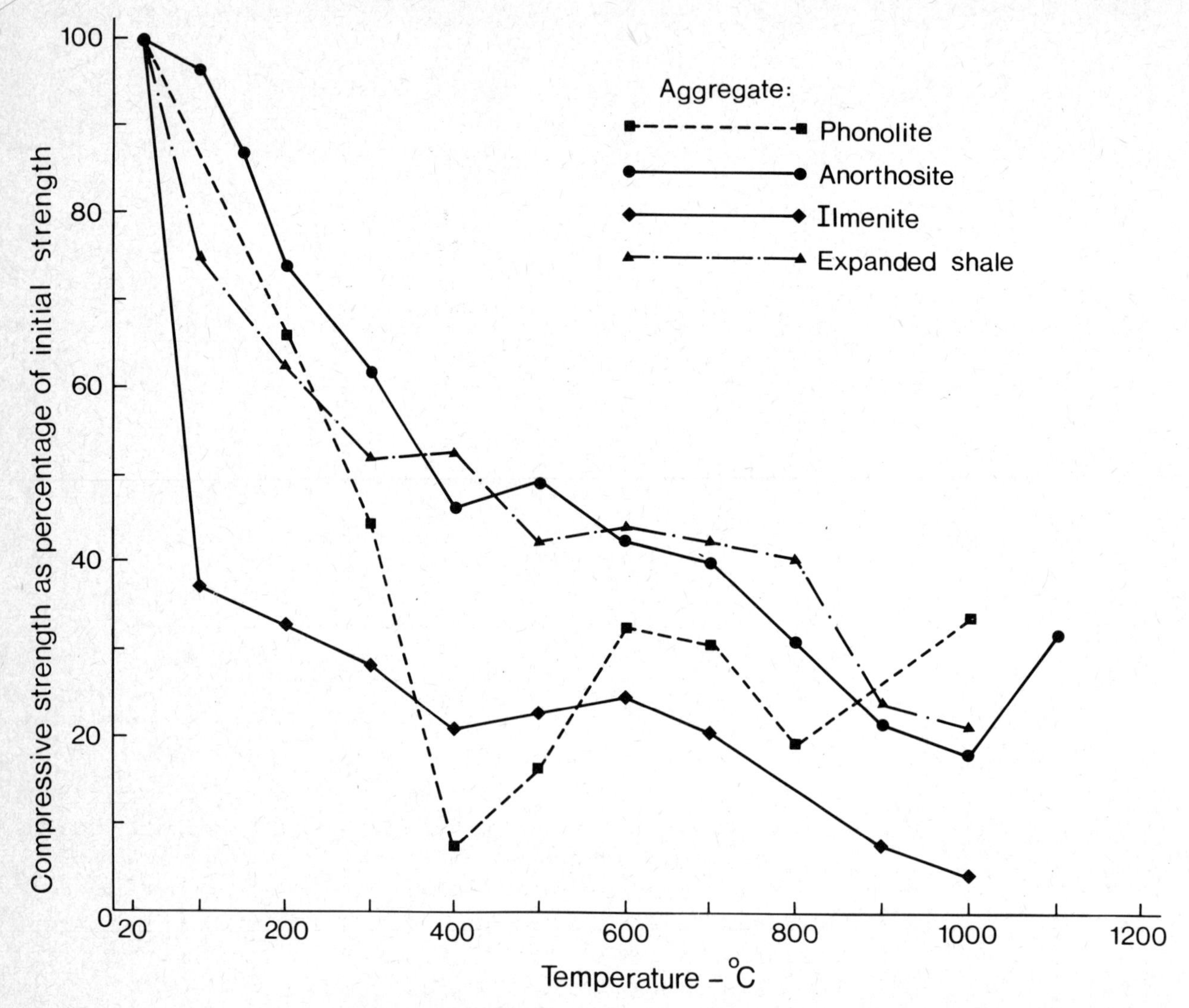

As a result, HAC concrete can withstand very high temperatures: with crushed firebrick aggregate up to about 1350°C, and with special aggregates up to 1600°C. HAC concrete is thus an excellent refractory material. This use was not appreciated when HAC was first developed but today it is for this feature that it is best known and most widely used.

White HAC

Even higher temperatures can be achieved using a white HAC (see Section 2.1) with special aggregates such as fused alumina: up to 1800°C can be resisted over prolonged periods of time.

Refractory HAC concrete can be brought up to service temperature as soon as it has hardened, i.e. it does not have to be pre-fired. On heating the first time, the concrete undergoes thermal expansion. This is balanced, however (with some excess) by the shrinkage of the cement paste, so that the nett contraction or expansion (depending on the aggregate) is low. The important point is that expansion joints are not needed. Butt joints are usually provided (at 1m to 2m); these open slightly on cooling below the service temperature.

Refractory use

In all cases, the choice of a suitable aggregate with appropriate heat-resisting properties is important. With lightweight aggregates, HAC can be used as insulating refractory concrete at temperatures up to around 950 °C.

The refractory uses are not structural but they are of interest. HAC is used in the construction of brick and tunnel kilns, for kiln linings, chimney linings and kiln doors, and in the construction of foundations for furnaces, coke ovens, and boiler settings. Special shapes can be readily moulded and monolithic construction is possible. Mortar can be shotcreted (gunited) to form refractory linings. The concrete can withstand a considerable thermal shock.

2.4 High early-strength

Strength development

Another useful property of HAC, again not realized when it was first developed, is its extremely high rate of gain of strength. It should be stressed that HAC is not quick-setting, its setting time being comparable with that of Portland cement (see Section 2.1). However, after HAC has set, it gains strength so rapidly that within 24 hours the compressive strength of concrete made with it can be as high as 90 per cent of the ultimate strength. Fig. 2.1 shows a typical strength-time curve.

Apart from the use in the precast concrete industry, which will be discussed later, many other advantages can be taken of this property of HAC. It can be used for all types of emergency repair work where the time factor is critical. Examples of this are: repair work on busy roads, floor and machine bases in factories, pipe jointing, and repair work on railways and airfields. HAC is used also for its rapid development of strength in sea defence work where parts of a structure may be accessible only between tides.

Setting properties

In connection with repair work, more should be said about the setting properties of HAC. When Portland cement and HAC are mixed there is a reduction in both the setting time and the strength compared with either component alone. For certain mixes of the two cements the setting time can be reduced to such an extent that it becomes almost instantaneous. This phenomenon is known as *flash setting*[2]. The relation between the setting time and the proportion of HAC to Portland cement is shown in Fig. 2.4 but this is only indicative of the behaviour. For any practical application, the actual proportions of cements to be used should be tried out. Advantage is sometimes taken of the flash setting of cement mixtures in repair work, for instance, when ingress of water cannot be stopped for more than a few minutes; mortar rather than concrete is usually employed.

The flash setting of Portland cement and HAC mixtures means that accidental contamination must not be permitted, and a scrupulous cleanliness of mixers and tools is necessary. Concretes made with the two cements can, however, be placed against one another, providing the older of the two is hard enough to resist intermixing at the interface. The required period may vary with temperature but is usually one day when HAC concrete is cast first and two days when Portland cement concrete is cast first.

Contaminants

This may be an appropriate place to mention that HAC must not be contaminated with materials that may interfere with its setting or hardening: Portland cement has already been mentioned but we should add plaster, lime or organic matter. The reactions of hydration are also sensitive to impurities in the mixing water, so that neither sea water nor impure water should be used with HAC.

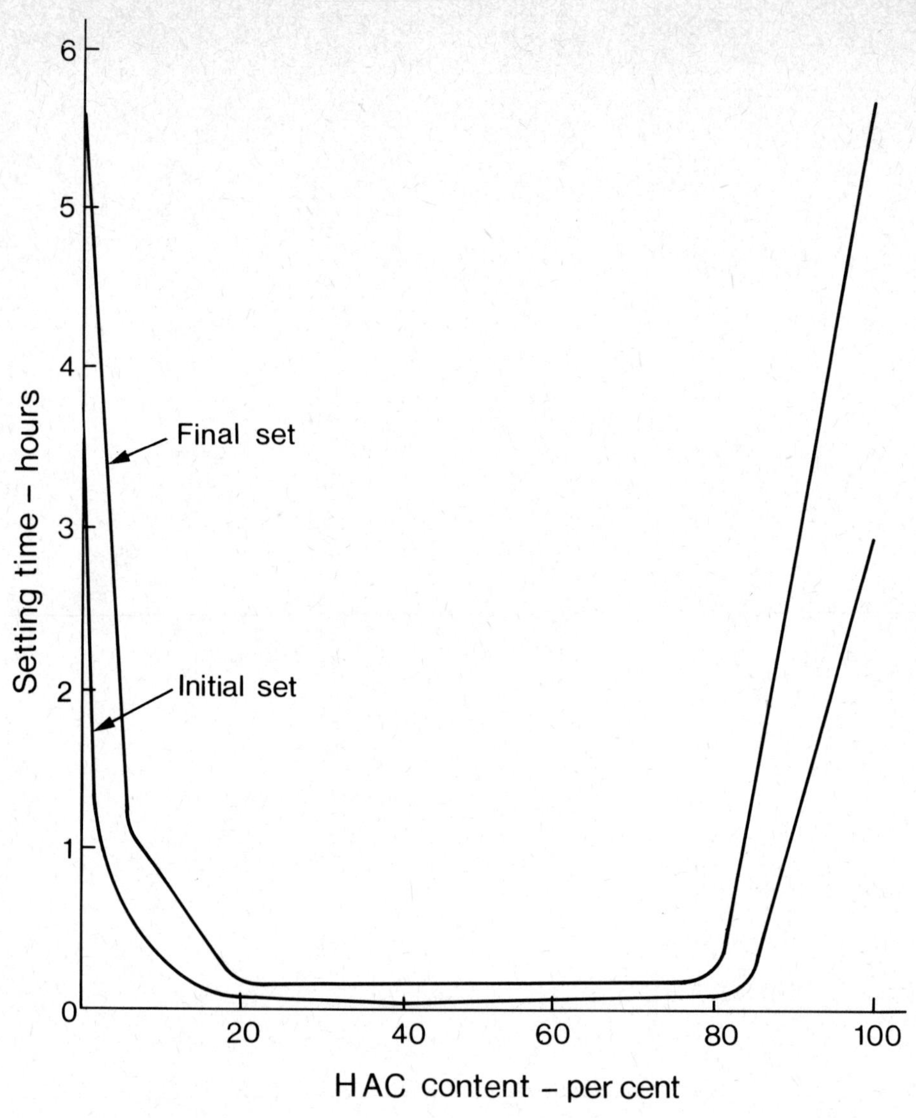

Figure 2.4 **Setting time of mixtures of HAC and Portland cement[2].**

Use in cold climate

As mentioned in Section 2.1, the total heat of hydration liberated by HAC is comparable to that of the same weight of ordinary or rapid hardening Portland cement. However, the rapid gain of strength of HAC means that the heat of hydration is released in a much shorter time than is the case with Portland cements. Under adiabatic conditions, the temperature rise within the cement is therefore much greater over a short period and this enables the cement to protect itself against severe frost. HAC is therefore employed in construction within cold storage plants and in mountainous and polar regions. However, when concreting under extremely cold conditions, a number of precautions must be taken: frozen aggregates should not be used, the temperature of the mixing water should not be close to freezing, and the mix should be protected against freezing during the first 4 to 6 hours, by which time the generation of heat starts. Provided such precautions are taken, high strengths are developed after a day or so. Fig. 2.5 shows some strength data for 70mm HAC concrete cubes made of a 1:6 mix with a water-cement ratio of 0.50. The mixing water had a temperature between 0°C and 4°C and the cement and aggregate were at the test temperature. Further tests at –6 °C have yielded a 24-hour strength of 28 N/mm^2[12].

Figure 2.5 **Strength development of HAC concrete at low temperatures[12].**

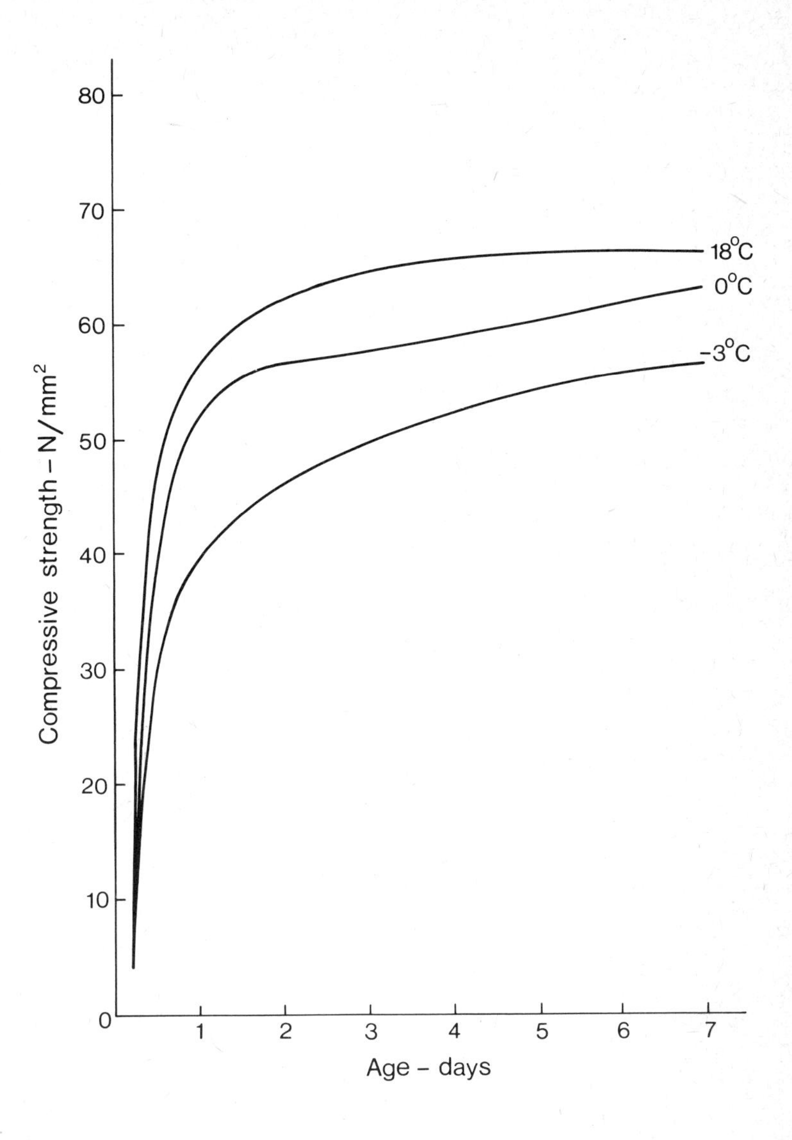

2.5 Use in structural precast concrete

As already mentioned, considerable use of the high-early-strength properties of HAC has been made in the precast concrete industry. In 1972 the industry consumed about one-quarter of the total volume of HAC used in the United Kingdom. The main use was in the production of prestressed concrete units, where the high early-strength allows the prestress to be transferred some 18 to 19 hours after casting. In precasting, from the economic point of view, it is important to have as quick a turn round on formwork as possible. This obviously allows the manufacturer to produce more units using a given number of forms, leads to the need of a smaller storage space for the units, and gives him an increased capital turnover. In the manufacture of precast pretensioned units, the forms can be removed only after the prestress has been transferred; this can be done when the compressive strength of the concrete has reached approximately 35 N/mm^2.

Obtaining high early strength

In order to achieve such a strength in as short a time as possible, the manufacturer has several options open to him. He can use ordinary Portland cement and cure the units at a high temperature by using saturated or unsaturated steam, or by heating the forms. He can adopt a similar procedure using rapid hardening Portland cement, which would be somewhat more expensive but would achieve a higher early strength. Using any of these methods it would be possible to release prestress probably within one to three days. A much shorter time would be obtained with ultra high-early-strength Portland cement (see Table 2.4).

Table 2.4

Typical values of strength of a 1:3 concrete made with ultra-high-early-strength Portland cement[124]

Age	*Compressive strength (N/mm²) at water-cement ratio of*		
	0.40	*0.45*	*0.50*
8 hours	12	10	7
16 hours	33	26	22
24 hours	39	34	30
28 days	59	57	52
1 year	62	59	57

The other choice open to the manufacturer is to use HAC. No high-temperature curing,and therefore no additional plant, is required and prestress can be released well within 24 hours. The choice, however, is not so simple as HAC costs three times as much, or more, as ordinary Portland cement. Moreover, the whole production must be geared to HAC. No Portland cement can be used on the same production line (unless everything is most stringently cleaned) because if the two cements are mixed there is a danger of flash setting (see Section 2.4).

Clearly, there are advantages in using HAC in precasting, and some manufacturers opted for HAC but others stuck to Portland cement and economically produced good quality precast, pretensioned concrete units for structural use. In each case, the choice, no doubt, was based on production and commercial factors, and possibly also on considerations of long-term performance of the product.

Commercial use

In fact, in the United Kingdom, two very large precast concrete manufacturers decided to use HAC, and between them they are probably responsible for the majority of prestressed HAC concrete units in use at the moment. Of their production, an estimated 90 per cent has been used in the beam and infill type floor and in roof construction, examples of which are shown in Fig. 2.6. The beam dimensions and prestressing details vary from beam type to beam type according to the span and loading requirements. The roof or floor screed normally used with these systems may or may not be of composite construction, as shown in Fig. 2.6. One large precast concrete manufacturer estimates that, using his system, short spans are particularly economical and more than one-half of his output has been for spans under 4m.

There is of course no reason why a structural designer should not use these units in systems of his own design. Thus they may be used as isolated (non-composite) members. This has been the case in a number of school buildings, including the Stepney school (see Chapter 7).

Figure 2.6 **Typical examples of floor and roof construction using HAC prestressed concrete beams:**

a) Composite floor construction
b) Beam and infill pot construction.

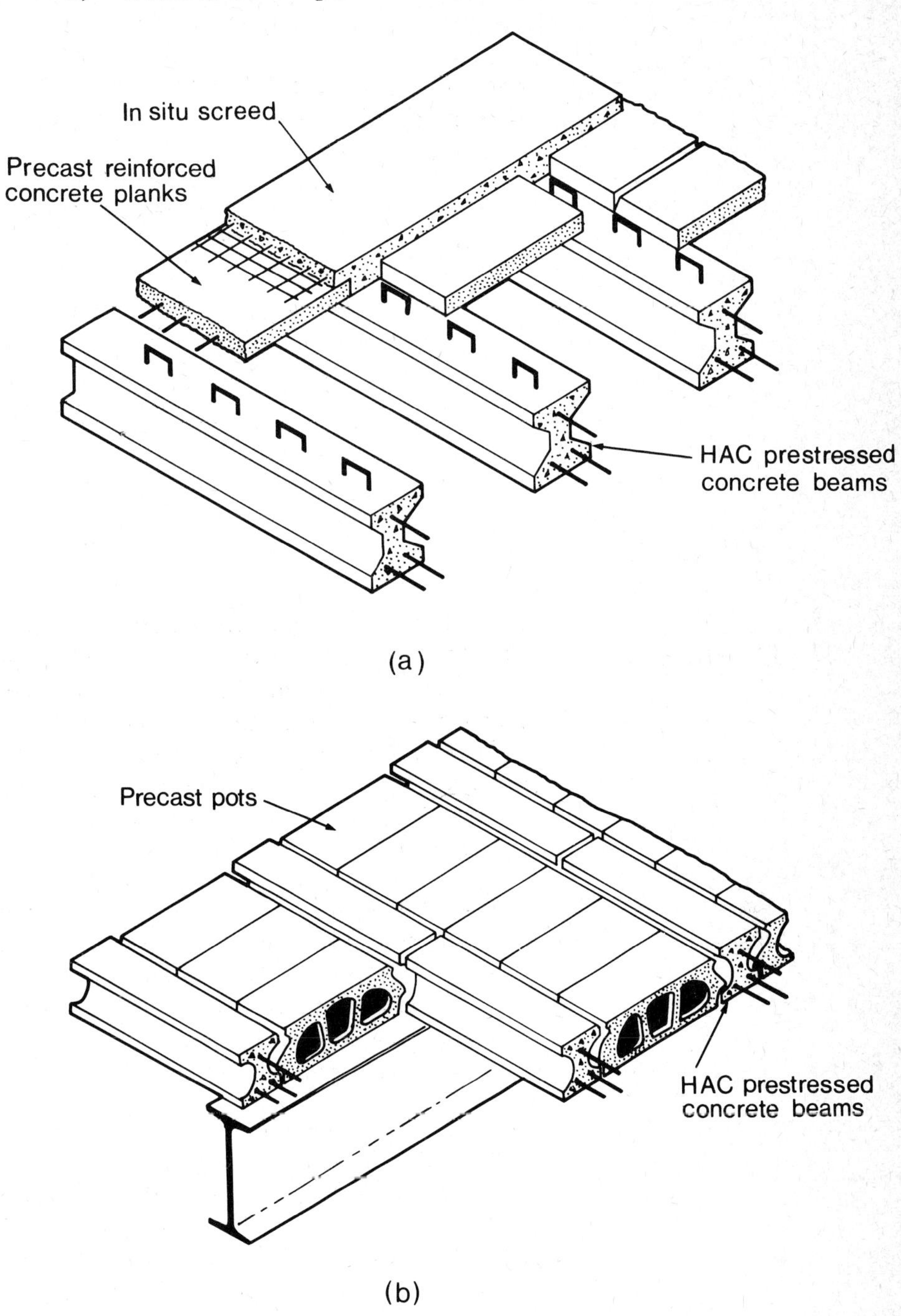

2.6 Use in foundations

The earlier discussion of the chemical resistance of HAC (Section 2.2) dealt with non-structural uses. But many structures have been built in soils containing a variety of harmful substances. Sulphate-bearing soils and soils that may previously have been on the site of a gas works or chemical works contain chemicals that could be deleterious to Portland cement. Depending on the concentration of these substances, the structural engineer has a number of choices open to him. It may be sufficient to use ordinary Portland cement in a concrete with an increased cement content. In some cases, it may be necessary to use a special type of Portland cement such as sulphate-resisting cement. However, under certain conditions, even sulphate-resisting cement may not be sufficient on its own, and the surface of the concrete may have to be protected with an impervious membrane. When conditions are so severe it may pay the designer to use a special type of cement, despite the fact that this is much more expensive than Portland cement. The two types of cement most commonly used under such conditions are HAC and supersulphated cement. Due to various production problems,the manufacture of supersulphated cement ceased in 1974, but there exist many buildings standing today on foundations made with supersulphated cement concrete. The same applies to buildings on HAC concrete foundations.

2.7 In situ work

HAC is only very rarely used for in situ structural work, except for emergency repairs and foundation construction mentioned earlier. The main reason for this is the high cost of HAC but there are also practical reasons related to workability and curing requirements. The water-cement ratios recommended in the Code of Practice CP 116: Part 2:1969[13] for the structural use of HAC are somewhat lower than those normally encountered with in situ work. Adequate compaction may therefore be very difficult to achieve with the normal methods of vibration available to the contractor. The code[13] also states that HAC concrete should be adequately cured: it should be continually sprayed with water for at least 24 hours after placing. Indeed, the atmospheric and temperature conditions in which the concrete is kept during its early age can have an important bearing on the strength at later life, much more so than with concrete made with Portland cements. Because of all this, the contractor may feel that such a high degree of control would not be practical under site conditions.

3 Conversion of High-Alumina Cement

3.1 What is conversion?

Since in the discussion of the 'HAC problem' the word conversion is repeatedly mentioned, it is important to explain fully the phenomenon and its consequences.

Hydration

First, a few words about the products of hydration of HAC. We mentioned in Section 1.2 that monocalcium aluminate $CaO.Al_2O_3$ is the main cementitious compound in HAC. In the presence of water, i.e. when the concrete has been placed in the formwork, at *ordinary temperatures,* chemical reactions take place, producing first alumina gel, and a few hours later $CaO.Al_2O_3 10H_2O$ is formed. Some $2CaO.Al_2O_3.8H_2O$ is also formed but is rarely present in excess of 1 or 2 per cent, the extent depending on the alkali content of the cement[2]. If, however, hydration proceeds at a higher temperature, say 45°C, the main product formed is $3CaO.Al_2O_3.6H_2O$.

The process of conversion

The detailed crystallographic structure of these compounds is not our concern but, since the terms are widely used, we may mention that both $CaO.Al_2O_3.10H_2O$ and $2CaO.Al_2O_3.8H_2O$ have pseudo-hexagonal symmetry while $3CaO.Al_2O_3.6H_2O$ is cubic*. Now, the important point is that at higher temperatures only the cubic form can exist; at room temperature either can, but the hexagonal crystals *spontaneously,* albeit slowly, convert to the cubic form. Application of heat speeds up the process. This then is conversion: an unavoidable change of one form of calcium aluminate hydrate to another, and it is only reasonable to add that such a change is not an uncommon phenomenon in nature.

Because they undergo a spontaneous change, the hexagonal crystals can be said to be metastable at room temperature, the final product of the reactions of hydration being the cubic form.

Before discussing the significance of conversion we should briefly describe the reaction. Conversion both of $CaO.Al_2O_3.10H_2O$ and of $2CaO.Al_2O_3.8H_2O$ proceeds direct; for instance:

$$3[CaO.Al_2O_3.10H_2O] \rightarrow 3CaO.Al_2O_3.6H_2O + 2[Al_2O_3.3H_2O] + 18H_2O$$

It should be noted that, although water appears as a product of the reaction, conversion can take place only in the presence of water and not in desiccated concrete. As far as neat cement paste is concerned, it has been found[14] that

*See Footnote to page 10

in sections thicker than 25 mm, the interior of the hydrating cement has an equivalent relative humidity of 100 per cent regardless of the environmental humidity so that conversion can take place. The influence of the ambient humidity is thus only on concrete near the surface.

The cubic product of conversion, $3CaO.Al_2O_3.6H_2O$, is stable in a solution of calcium hydroxide ($Ca(OH)_2$) at 25°C but reacts with a mixed $Ca(OH)_2$-$CaSO_4$ solution to form $3CaO.Al_2O_3.3CaSO_4.31H_2O$ both at 25°C and at higher temperatures[15].

Loss of strength

The chemical changes described above do not make it in the least clear why conversion may be linked to a loss of strength of concrete. The matter has not been resolved fully but the likely explanation is in terms of the densification of the calcium aluminate hydrates. The densities of the compounds involved are believed to be as follows[2,16]:

$CaO.Al_2O_3.10H_2O$	1.72 to 1.78 g/ml
$2CaO.Al_2O_3.8H_2O$	1.95 g/ml
$3CaO.Al_2O_3.6H_2O$	2.52 to 2.53 g/ml
$Al_2O_3.3H_2O$	about 2.40 g/ml

Thus, it can be seen that, even though $Al_2O_3.3H_2O$ occupies some space, a change from the hexagonal form to the cubic under conditions such that the overall dimensions of the body are constant (as is the case in set cement paste) results in an increase in the porosity of the paste.

Let us consider neat cement paste in which the reaction of conversion is

$$3[CaO.Al_2O_3.10H_2O] \rightarrow 3CaO.Al_2O_3.6H_2O + 2[Al_2O_3.3H_2O] + 18H_2O$$

This means that the corresponding weights involved are in the ratio

$$1014 \rightarrow 378 + 312 + 324$$

Dividing each weight by the appropriate density, we obtain the ratio for the volumes of the compounds involved[16], viz.

$$1 \rightarrow 0.261 + 0.228 + 0.568$$

Thus 1 ml of $CaO.Al_2O_3.10H_2O$ yields 0.489 ml of solid hydrates and 0.568 ml of water (or voids)[16]. This of course would apply only if the starting point were a solid mass of $CaO.Al_2O_3.10H_2O$. In the case of concrete the relative volume of voids would be proportionately lower.

Increase in porosity

Since, as mentioned earlier, the explanation of the loss of strength in terms of porosity is, or at least was, not unchallenged, further evidence should be mentioned. Measurements of gross density of saturated specimens[17] and of dry-stored specimens[18,19] have confirmed an increase in porosity on conversion. Tests[2] have shown that cements hydrated to the extent of about 90 per cent at 18 °C and at 45 °C, and then dried over calcium chloride, had densities of 2.11 and 2.64 g/ml respectively. This difference would indicate a porosity of 20 per cent in the converted *paste*.

Measurements[20] of ultrasonic-pulse velocity through *concrete* yielded a value of the velocity through converted concrete equal to about 0.85 of the velocity through the unconverted concrete near a cool surface. This ratio corresponds to a porosity of concrete of not less than 10 per cent.

The difference between the values of porosity for neat cement paste and for concrete clearly reflects the fact that, in the latter, the *gross* volume of hydrated cement paste occupies only one-fifth of the total volume. Thus the porosity is lower but still important. Fig. 3.1 shows the relation between porosity and strength in a typical concrete (made with HAC or Portland

Figure 3.1 **Relation between the volume of voids and reduction in strength in concrete.**

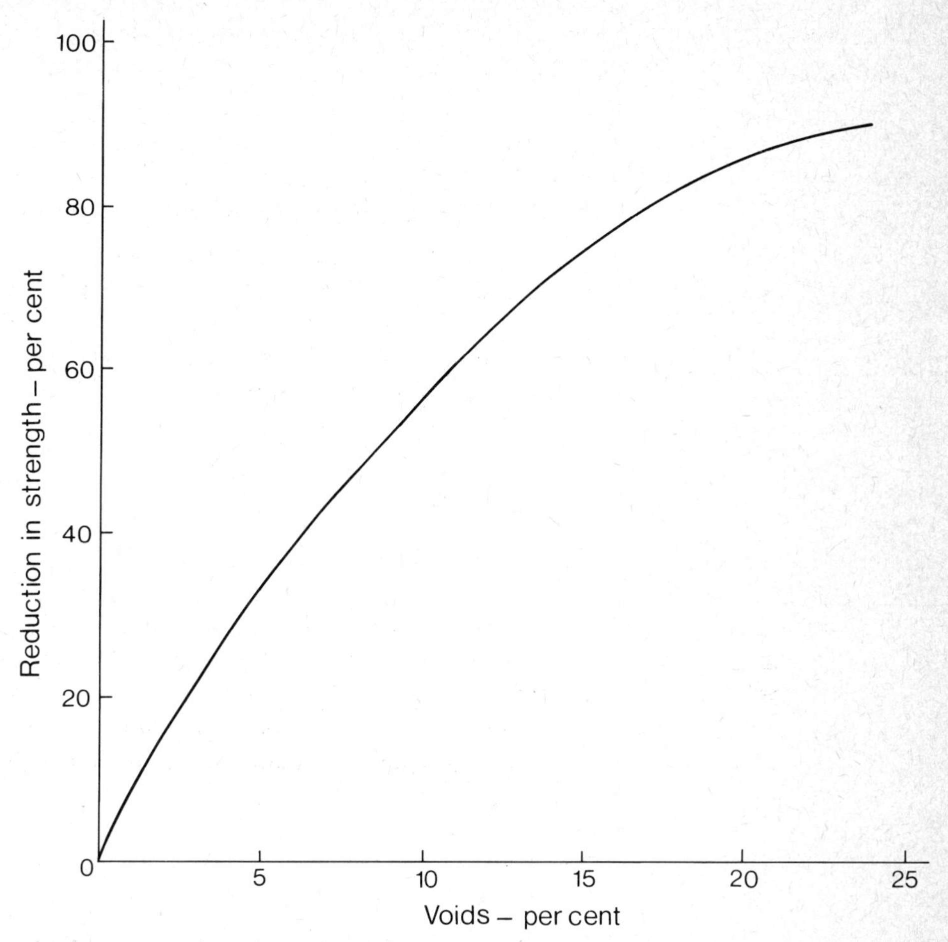

cement), and we can see that a porosity of 7 per cent results in a loss of 50 per cent of strength compared with non-porous concrete. It is only fair to add that all pores, whatever their origin, affect the strength of concrete. What we are talking about here are the *additional* pores due to conversion, and therefore a loss of stength compared with the initial, unconverted state.

In 1960, we obtained some interesting measurements[17] on the increase in weight of HAC concrete stored in cold and in hot water. Fig. 3.2 shows that the weight at 40°C increased continuously with time, but at a decreasing rate, as would be expected in the case of continuing hydration. The rate of increase, as well as its magnitude, is greater the richer the mix because the richer the mix the larger the volume of the paste and the greater the amount of water it can hold. The increase in weight was found to be higher at higher temperatures, the explanation being in terms of the higher porosity of the concrete in which the HAC was converted. Fig. 3.3 shows the percentage increase in weight after 200 days plotted against the volumetric content of cement paste in the mix for storage at 18°C and 40°C: the two increase together. However, the slopes of the curves flatten out somewhat because mixes with a higher cement paste content had a lower water-cement ratio and therefore a lower initial porosity. The difference in the ordinates of the two lines of Fig. 3.3 for a given abscissa indicates the additional water held by the converted paste. This water occupies the volume of pores produced as a result of conversion, and corresponds to the change in the specific gravity of the calcium aluminate hydrate. We should note that conversion increases the air porosity of HAC concrete by a factor of about thirty.

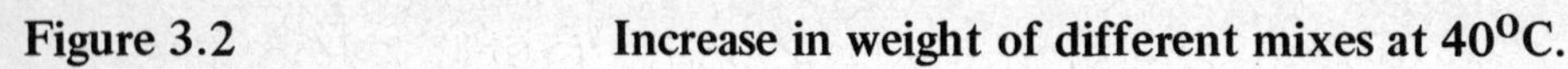
Figure 3.2 **Increase in weight of different mixes at 40°C.**

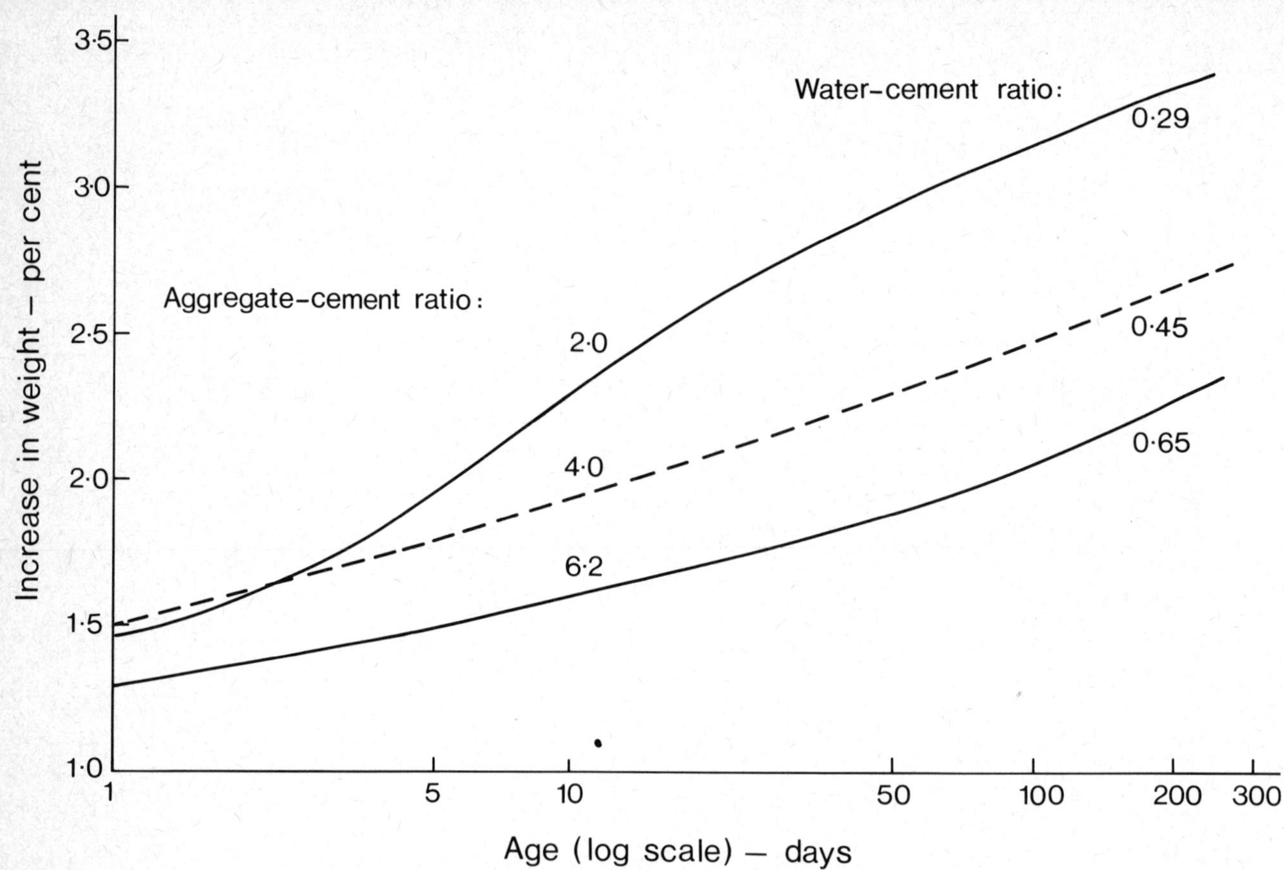

The suggestion that the loss of strength is due to an increase in the porosity of the cement paste has some support also from the observation that in some cases of very slow conversion, when continuing hydration of hitherto unhydrated cement resulted in the formation of products occupying some of the newly-formed pores, neither the porosity substantially increased nor was the strength greatly reduced. Research on neat cement paste led to the suggestion that at low water-cement ratios, the products of hydration are non-porous[14]. The explanation probably is that conversion is accompanied by a simultaneous hydration of the hitherto unhydrated cement by means of the water released by the reactions of hydration. Since the data were obtained for neat cement paste, it is not obvious that exactly the same conditions would be obtained in concrete. This topic is discussed further in Section 4.2 but is of limited practical interest.

Another explanation[74] of the relatively low porosity of pastes with low water-cement ratios is that the crystal size of the converted aluminates is lower than at higher water-cement ratios. For instance, at a water-cement ratio of 0.25, the crystal size is estimated to be 0.5μm while at a water-cement ratio of 0.75 the size is 2.5μm. Hence, packing of the crystals is affected. The much larger size of the converted aluminates compared with the unconverted form was demonstrated in 1971[125].

Reverting to the causes of conversion, it is only fair to add that there are other possible reasons why conversion of the calcium aluminates may lead to a loss of strength. Generally, the dissociation of a hydrate results in a decrease in strength and in the modulus of elasticity, but it is not known whether this is due to the dehydration itself or to the formation of the

Figure 3.3

Relation between volumetric content of cement paste and increase in weight after 200 days for storage in water at 18°C and at 40°C.

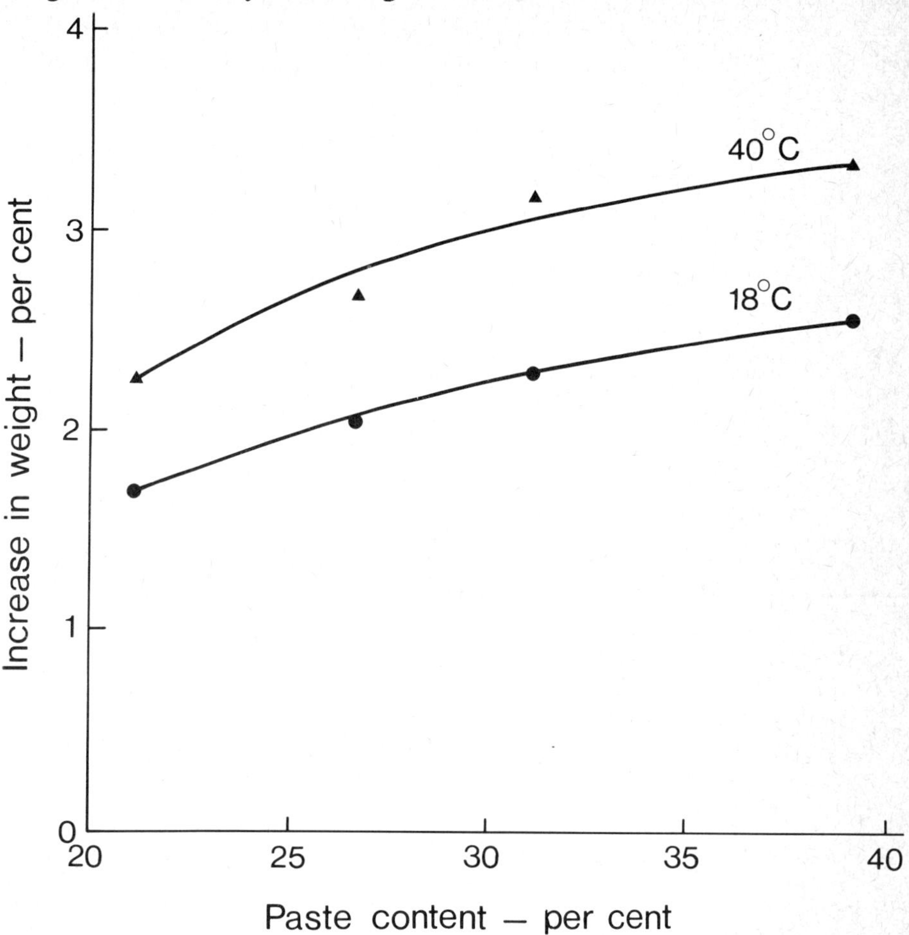

products of the reaction[21]. Further important factors in strength are the degree of perfection of the crystals and the grain size; it is also possible that tension[22] induced by movement of water or its vapour at the crystal-size level tends to cause some disruption. It has been suggested that aging of microcrystalline Al_2O_3 . aq to the macrocrystalline gibbsite (which occurs at the same time as the conversion of the calcium aluminates to the cubic form) is the cause of a decrease in strength. The argument is based on the belief that the colloidal or gelatinous hydrated alumina is as important to the strength of HAC as the colloidal or gelatinous hydrated calcium silicates are in Portland cement[23].

Influence of water

Influence of water *per se* on the loss of strength has also been mentioned as a possibility. A large decrease was observed[16] in the compressive strength of mortar consisting of the hydrates $3CaO.Al_2O_3.6H_2O$ and $Al_2O_3.3H_2O$ when soaked in water. It has been suggested that such a decrease, of the order of 30 to 40 per cent, may also be provoked by the water produced in the course of conversion, and not only by water imbibed from outside[16]. This role of water may be a secondary factor in the influence of conversion on the loss of strength. Whatever the exact mechanism causing the loss of strength, there is no doubt that it is the conversion that leads to a loss of strength.

Change in colour

The loss is often accompanied by a change in the colour of the cement paste but this change is now known not to affect the strength. The change is from the characteristic blackish-grey to brown in the case of cements with a high ferrous oxide content or to yellow when the oxide content is low, and is due to the oxidation of the ferrous oxide to the ferric form. (The presence of iron in ferrous or ferric form depends on the extent of the oxidizing atmosphere during the manufacture of cement.) Suggestions that there is a connexion between the change in colour and the loss of strength have been made from time to time, but cements with a ferrous oxide content varying between 1 and 10 per cent[24] and a total iron oxide content between 1.4 and nearly 19 per cent[25] have all been found to lose strength, and so has iron-free neat $CaO.Al_2O_3$ when cured at higher temperatures[24]. The increase in porosity on conversion, however, facilitates the oxidation of the ferrous compounds, and this is why conversion and change in colour occur together, but concrete which is porous because of bad mix proportions has been shown to change colour without conversion taking place[26].

Effect of temperature

The temperature at which the cement is formed does not affect the pattern of conversion as cements manufactured by the usual method of fusion, as well as those made by sintering (see Table 1.3)[25] all behave in a similar manner. Some Russian tests[61] also show a loss of strength on conversion. The loss of strength takes place regardless of the oxide composition of cement, as shown by tests[21] covering the range:

Al_2O_3	37 to 81 per cent
CaO	17 to 39 per cent
SiO_2	0.0 to 10 per cent
Fe_2O_3 and FeO	0.3 to 15.5 per cent

By comparison, the usual commercial HAC contains about 35 to 44 per cent of each of Al_2O_3 and CaO. White HAC has 60 to 70 per cent of Al_2O_3.

In view of all this, the suggestion that some, notably foreign, cements are liable to conversion, while the British-made cement is different does not bear scrutiny.

Degree of conversion

The degree of conversion is estimated from the percentage of $3CaO.Al_2O_3.6H_2O$ present as a proportion of the sum of the cubic and hexagonal hydrates together, viz.

degree of conversion (per cent)

$$= \frac{\text{weight of } 3CaO.Al_2O_3.6H_2O}{\text{weight of } 3CaO.Al_2O_3.6H_2O + \text{weight of } CaO.Al_2O_3.10H_2O} \times 100$$

A more practical expression gives the degree of conversion in terms of the weight of gibbsite ($Al_2O_3.3H_2O$) which is formed by the reaction of conversion. Thus

degree of conversion (per cent)

$$= \frac{\text{weight of } Al_2O_3.3H_2O}{\text{weight of } Al_2O_3.3H_2O + \text{weight of } CaO.Al_2O_3.10H_2O} \times 100$$

The two expressions do not give exactly the same result because of the difference in the molecular weights of $Al_2O_3.3H_2O$ and $3CaO.Al_2O_3.6H_2O$ but at high degrees of conversion the difference is not significant. Measurement of conversion on the basis of gibbsite is considered preferable because it is not affected by carbonation.

Most laboratories report the result to the nearest 5 per cent.

3.2 Conditions under which conversion occurs

The earliest work on the loss of strength of HAC due to what became known as conversion was done at the Building Research Station as far back as 1933. In February 1951, the Station issued Digest No. 27[27] in which it was stated that *"High alumina cement concrete should not be used in places where it will be both moist and at a temperature of about 85 °F (29 °C) or higher. These conditions acting together cause a loss in strength, whether they occur early or late in the life of the concrete. The residual strength is still high enough to be adequate for structural purposes but the general quality of the concrete is reduced; in particular, it seems to be much more readily attacked by salt solutions or water which can cause complete disintegration. Apart from certain natural and industrial conditions in which these two factors of moisture and elevated temperature occur, the rapid heat evolution of the cement itself during setting and hardening can cause a temperature rise inside the concrete mass; as the concrete is still moist at this stage, the two conditions leading to low strength can be self-produced. To avoid them certain simple, but important, precautions in placing the concrete are*

Figure 3.4 **Compressive strength of 1:2:4 HAC concrete with a water-cement ratio of 0.6 stored at 16°C and 35°C[27].**

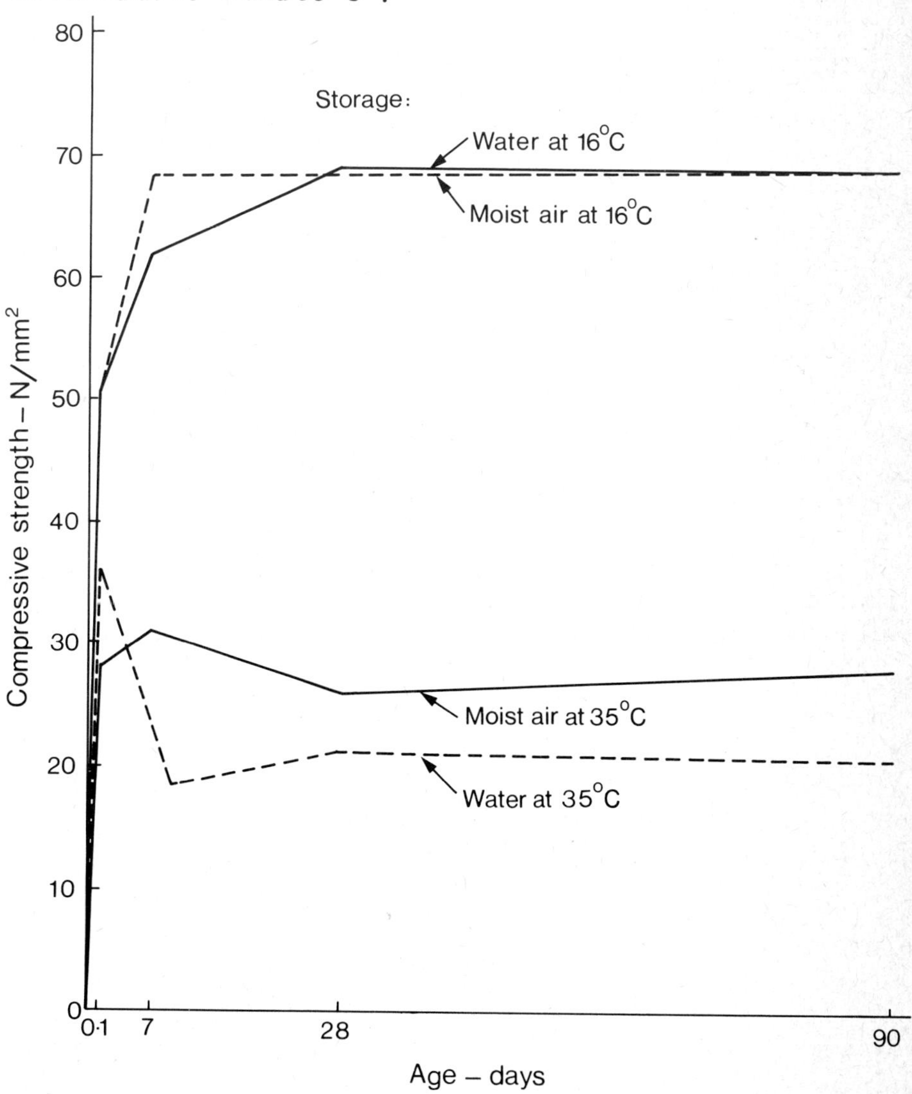

necessary. They are:–

(i) *Avoid mixing and placing operations when the atmospheric temperature is likely to remain above 85°F (29°C) for any length of time.*

(ii) *Restrict the maximum thickness of concrete placed in one operation to 1 ft 6 in (450 mm).*

(iii) *Strip the formwork as soon as possible and spray the concrete with water.*

High alumina cement concrete which has been affected by becoming too warm usually has a noticeably brown colour. The external appearance is not, however, an infallible guide because, when the concrete has become too hot through its own heat evolution during setting and hardening, the effect of this may be confined to the interior and not appear on the surface. The latter then shows no colour change and seems strong and hard though the interior has been affected."

The Digest[27] then gives figures on the effect of temperature, which are reproduced in Figs. 3.4 and 3.5. It is important to note that the Digest recommended the use of concrete mixes not richer than 1:5, nor leaner than about 1:7, and suggested that *"about 5½ gallons of water to each 1 cwt of cement is the best* (i.e. a water-cement ratio of 0.49), *but with normal 1:2:4 concrete it should never be less than 4½ gallons to 1 cwt* (i.e. a water-cement ratio of 0.40)*"*.

This was an early warning about conversion, but even earlier information exists. As far back as 1933, there was a laboratory investigation at the Building Research Station[28] and a further report in 1940[24]. Cases of deterioration of HAC under hot conditions were reported in 1945[29], 1949[30], and 1952[4]. The effect of an excessive rise in temperature during setting was reported in 1930[31], and in 1957 we published a report[32] on the

Figure 3.5 **Compressive strength of 1:2:4 HAC concrete with a water-cement ratio of 0.6 after storage at:-**

1. 24 hours at 18°C, thereafter 38°C.
2. 7 days at 18°C, thereafter 38°C
3. 28 days at 18°C, thereafter 38°C
4. 3 months at 18°C, thereafter 38°C[27].

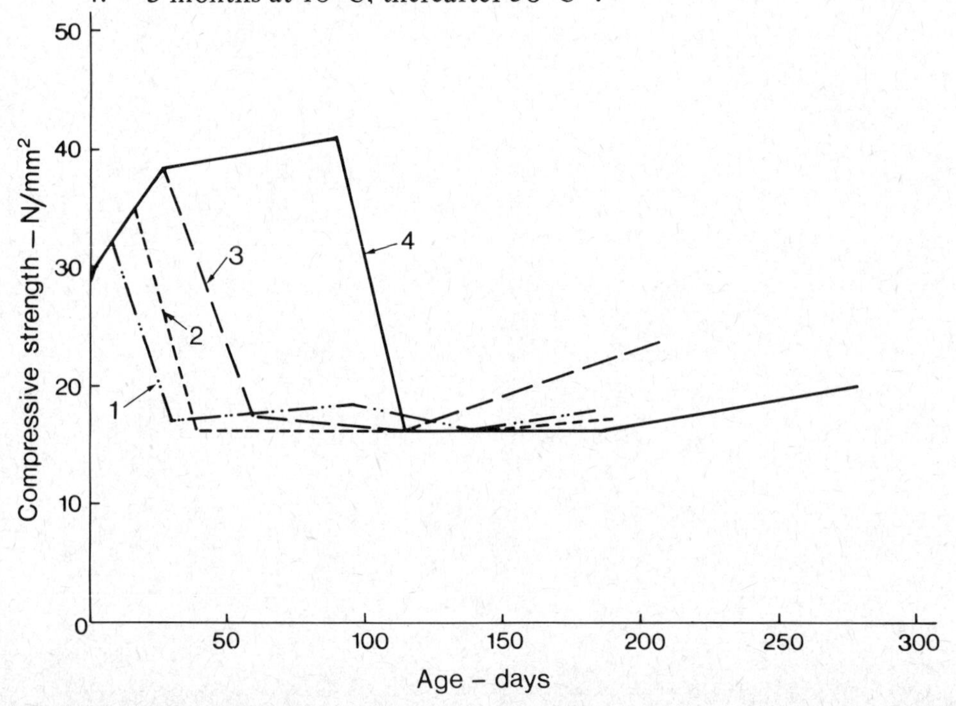

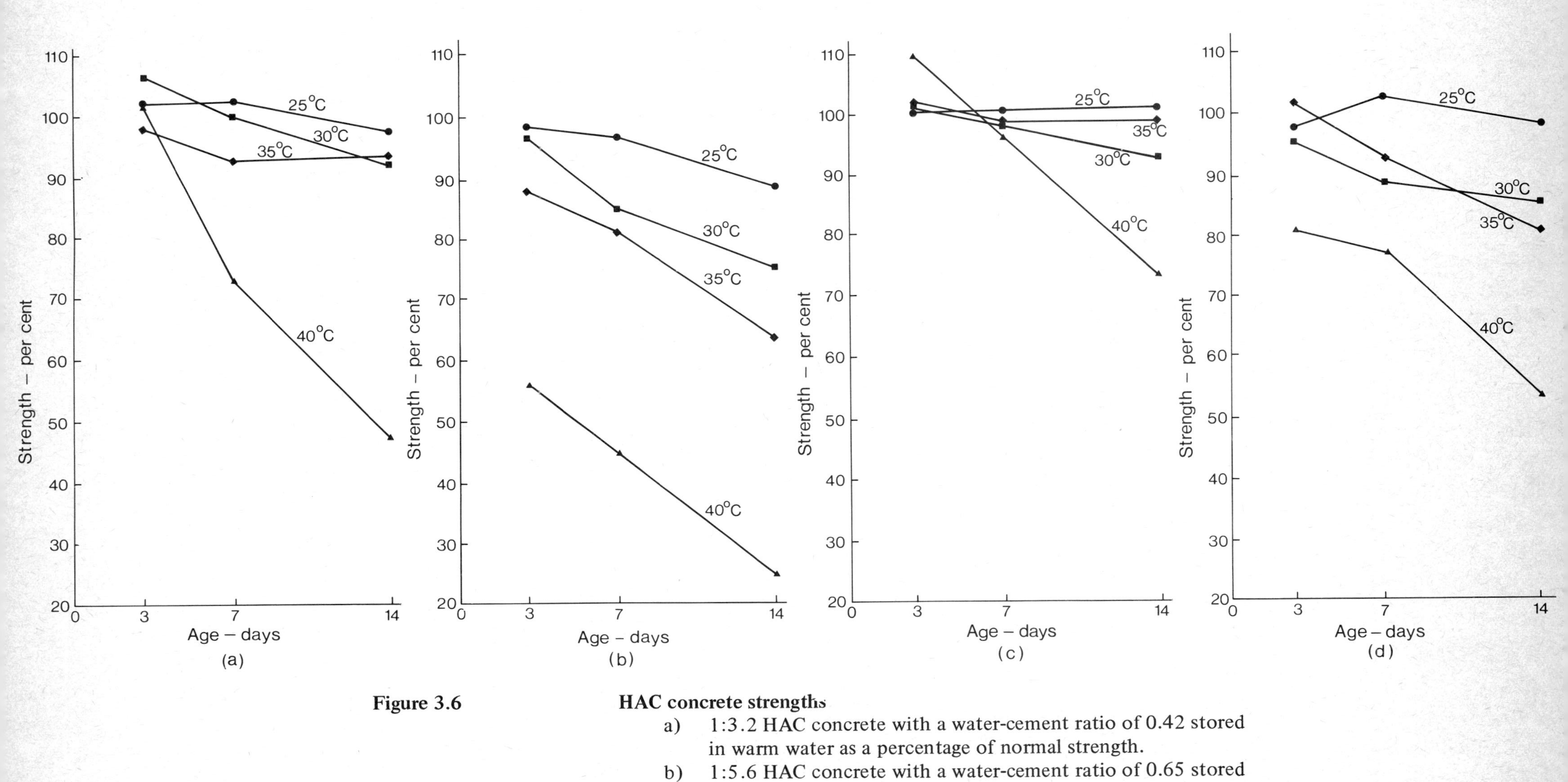

Figure 3.6 **HAC concrete strengths**

a) 1:3.2 HAC concrete with a water-cement ratio of 0.42 stored in warm water as a percentage of normal strength.
b) 1:5.6 HAC concrete with a water-cement ratio of 0.65 stored in warm water as a percentage of normal strength.
c) 1:3.2 HAC concrete with a water-cement ratio of 0.42 stored above warm water as a percentage of normal strength.
d) 1:5.6 HAC concrete with a water-cement ratio of 0.65 stored above warm water as a percentage of normal strength.

rapid loss of strength caused by radiant heat. Since, however, little was known at the time about the behaviour of HAC concrete subjected to no more than warm (little above normal) temperature, we investigated this problem in 1958, including the case when concrete is subjected to warm temperatures from time to time only[33]. In that investigation, cements more modern than pre-war, and therefore of higher strength were used.

Effect of storage conditions

The results of the investigation are summarized in Fig. 3.6, which shows clearly how the strength of the concretes investigated is affected by the temperature of storage. In the first few days, warm storage may actually result in an increase in strength (owing to accelerated reactions of hydration) as compared with storage in water at room temperature, but subsequently the strength falls off with time and generally more rapidly the higher the temperature. The effect is larger in the case of a 1:5.6 mix with a water-cement ratio of 0.65 than of a 1:3.2 mix with a water-cement ratio of 0.42.

Storage above hot water (Fig. 3.6 c and d) was found to have a similar but less marked effect, probably because the relative humidity fluctuated. Nevertheless, the effect of such storage on strength is of considerable interest since many structures are exposed to warm atmosphere above water.

The effect on HAC concrete of storage alternating between water at 35°C and water at room temperature for equal periods of time is shown in Fig. 3.7. Fig. 3.8 shows the same data plotted all together regardless of the period of alternation: the gradual loss of strength is unmistakable. It seems, however, that periodic cold storage results in a slight recovery of strength. This is apparent for the 14-day and to some extent also for the 7-day periods of alternation. This subject is discussed in Section 4.2.

For the shorter periods, the loss of strength is somewhat lower than for similar continuously warm-stored specimens but the effect of individual cold periods is less apparent; this is probably because of the time required for the change in temperature to take place throughout the specimen. Indeed, it is probable that for the short periods of alternation the interior of the specimen never reaches either of the extremes of the temperature of the storage water, but is subjected to a nearly constant temperature somewhere between the two external extremes.

Since the rate of strength recovery is generally less than the rate of loss there is little indication that the effects of warm storage can be undone by subsequent cooler storage. The rate of loss of strength is lower the older the concrete when first subjected to warm conditions, but the ultimate loss is the same. It may be noted that the loss shown in Fig. 3.7 would be higher for mixes with a higher water-cement ratio than for a mix with a water-cement ratio of 0.42.

From this investigation it was concluded that storage of HAC concrete in humid conditions at temperatures even little above normal results in a definite gradual loss of strength. This loss is greater the higher the temperature of storage, the higher the humidity, and the lower the concrete strength. Periods of cold storage appear to abate this loss but do not produce a recovery of equal magnitude. Since short periods of exposure to warm and humid conditions can often occur, the data reported in 1958 were thought to be of interest to the construction industry, and the paper[33] ended by a warning that "*with the modern tendency to higher working stresses, particularly in prestressed concrete, the deterioration described cannot be always ignored.*"

Figure 3.7 **Strength of a 1:3.2 HAC concrete with a water-cement ratio of 0.42 subjected to alternating storage as a percentage of normal strength.**

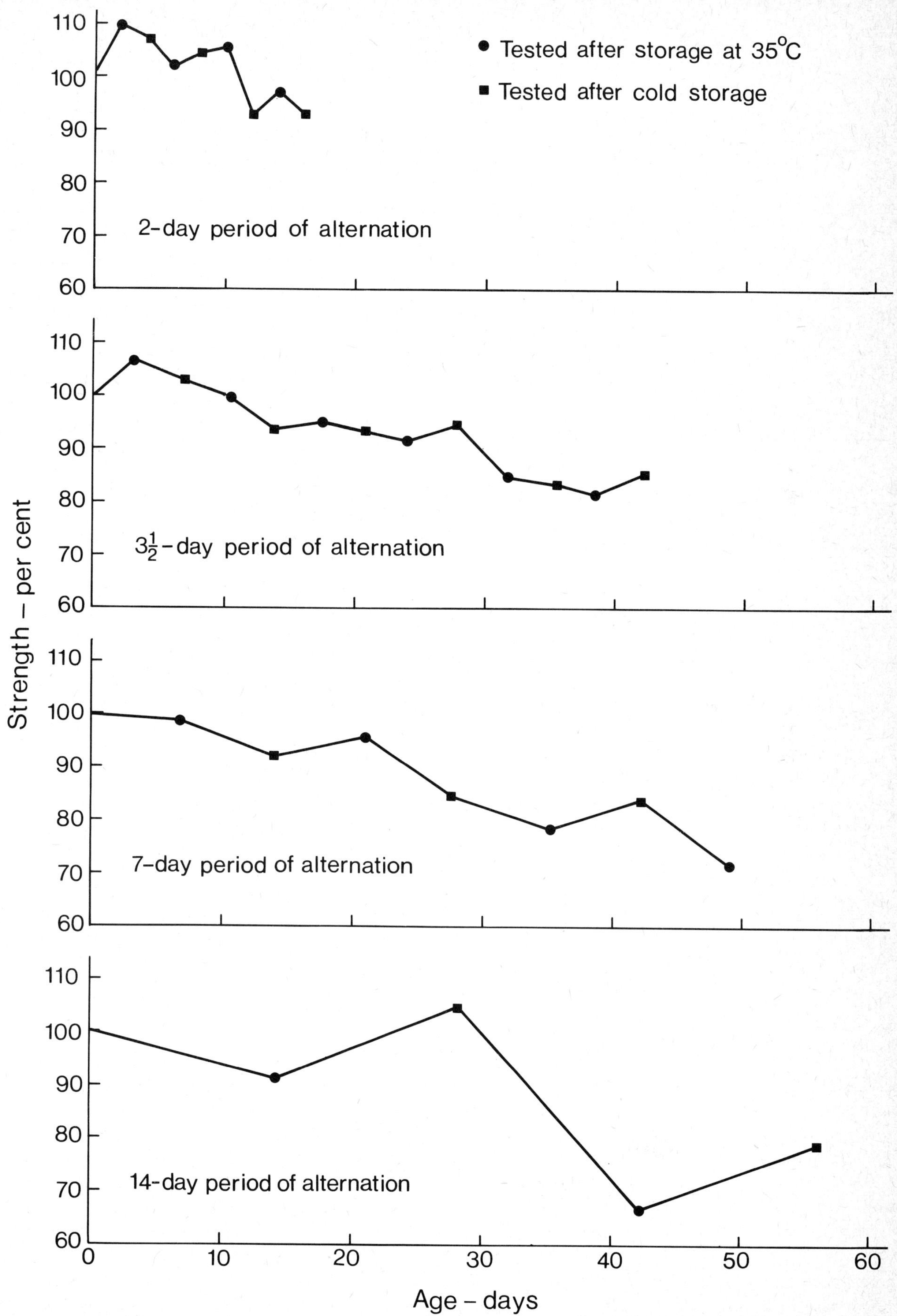

Figure 3.8 **Strength of HAC concretes (1:3.2 mix with a water-cement ratio of 0.42 and 1:5.6 mix with a water-cement ratio of 0.65) subjected to continuous and intermittent warm storage as a percentage of normal strength.**

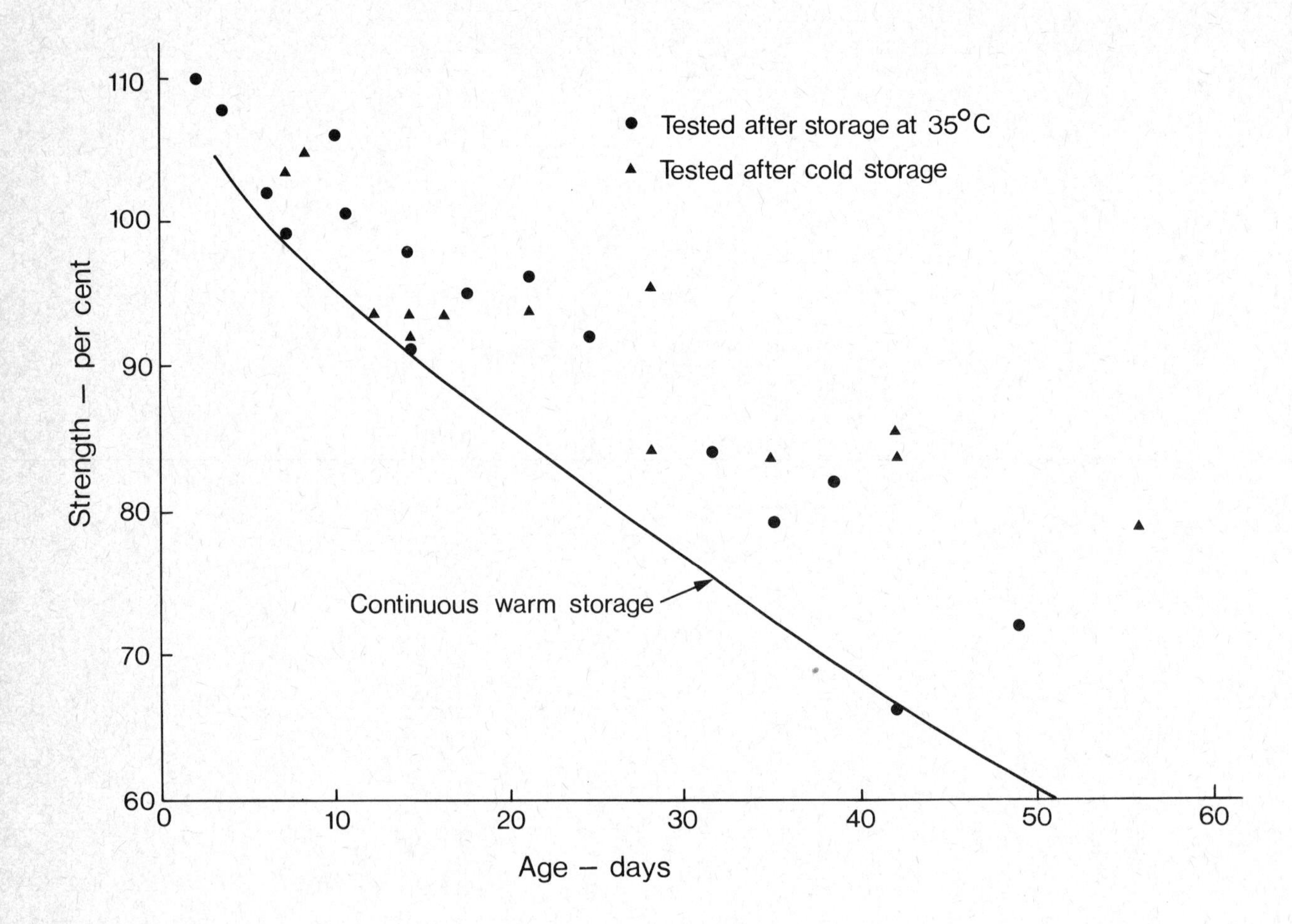

3.3 Factors affecting conversion

Temperature

The rate of conversion depends on temperature. The relation [14] between the time necessary for one-half of the $CaO.Al_2O_3.10H_2O$ to convert and the temperature of storage of 13 mm cubes of neat cement paste with a water-cement ratio of 0.26 is shown in Fig. 3.9 [14]. It is likely that for the more porous concretes of practical mix proportions, the periods are much shorter as full conversion has been observed after some twenty years at 20°C or thereabouts. Thus data on neat cement pastes with very low water-cement ratios should be used circumspectly, but they are nevertheless of scientific interest.

The same applies to the relation between the strength of neat HAC paste cubes, 13 mm in size, at the time when half-conversion (50 per cent) has taken place, and the temperature of storage [14]. This is shown in Fig. 3.10 and suggests that the higher the temperature the lower the strength at the same degree of conversion. The tendency is clear but again the situation may not apply in the case of full-size concrete members.

It may be interesting to note what happens at highly elevated temperatures. Experiments [14] on neat HAC paste with a water-cement ratio of 0.26 showed that the reactions of conversion at warm temperatures are followed,

Figure 3.9 **Time for half-conversion of neat HAC pastes (13 mm cubes) cured at various temperatures[14].**

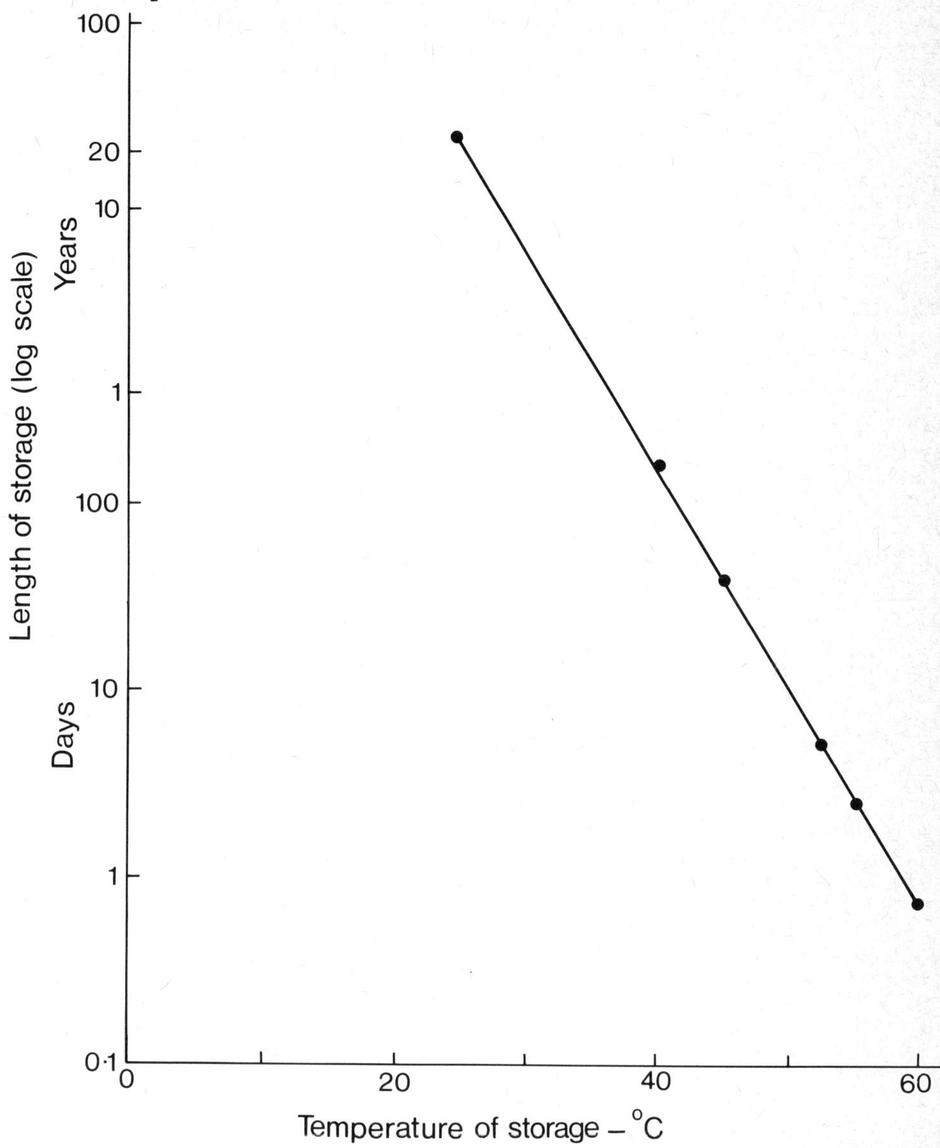

Table 3.1 **Influence of water-cement ratio on loss of strength on conversion**

Cement	*Water-cement ratio*	*Aggregate-cement ratio**	*100-day strength at 18°C† (N/mm²)*	*Strength of converted concrete as a percentage of strength at 18°C*
A	0.29	2.0	91.0	62
	0.35	3.0	84.4	61
	0.45	4.0	72.1	26
	0.65	6.2	42.8	12
B	0.30	2.1	92.4	63
	0.35	3.0	80.7	60
	0.45	4.0	68.6	43
	0.65	6.2	37.2	30
	0.75	7.2	24.5	29

* Maximum size of aggregate 10mm

† 76mm cubes

Figure 3.10 **Strength of neat HAC pastes (13mm cubes) after half-conversion[14].**

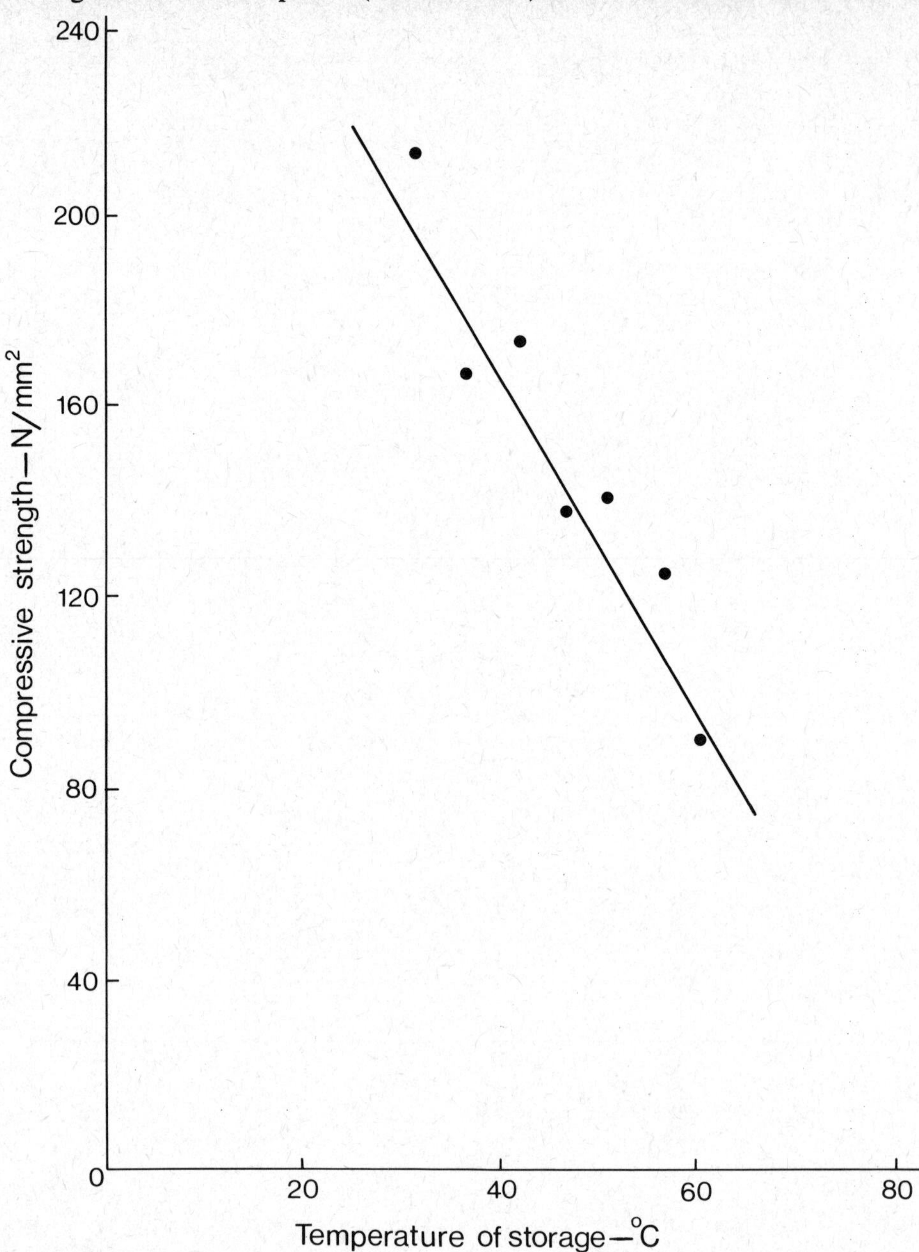

when the temperature is raised to 200°C, by a decomposition of $3CaO.Al_2O_3.6H_2O$ and a formation of a mixture of $12CaO.7Al_2O_3$, $Ca(OH)_2$, and of amorphous Al_2O_3. This is accompanied by a further decrease in strength. On further heating, $Ca(OH)_2$ dehydrates without a loss of strength until, between 900 °C and 1000 °C, some of the $12CaO.7Al_2O_3$ reacts with Al_2O_3 to produce $CaO.Al_2O_3$; this reaction is accompanied by a further loss of strength.

The results so far considered indicate that concretes and mortars of any mix proportions lose strength when exposed to a higher temperature, and the general pattern of the loss–time curves is similar in all cases. However, the degree of loss is a function of the water-cement ratio of the mix, as shown in Fig. 3.11. The mix proportions and percentage loss are given in Table 3.1. It is clear that the loss, either in Newtons per square millimetre or as a fraction of the strength of the cold-cured concrete, is smaller in mixes with low water-cement ratios than in mixes with high water-cement ratios.

Figure 3.11 **Influence of the water-cement ratio on the strength of HAC concrete cured in water at 18°C and at 40°C for 100 days.**

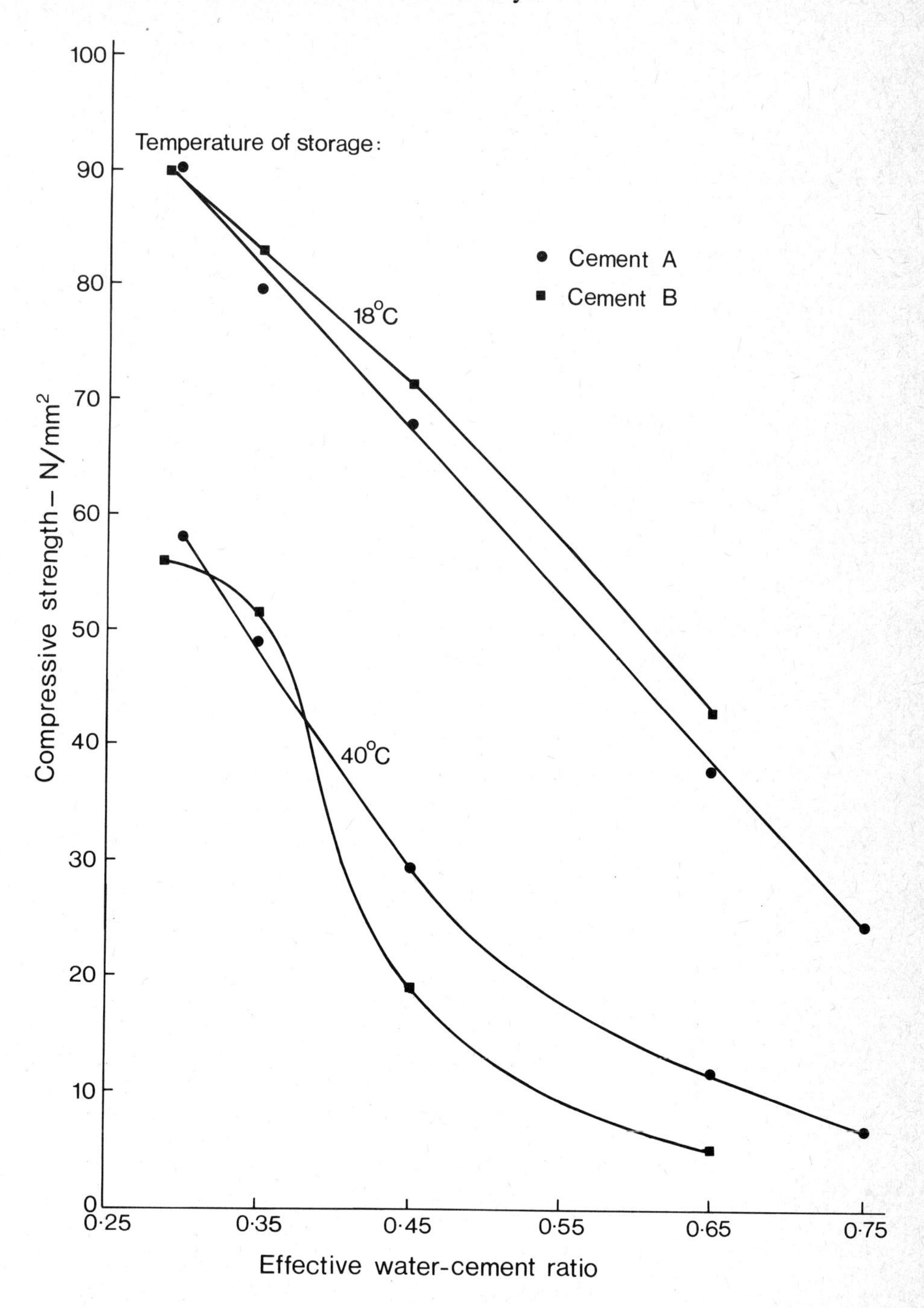

Water-cement ratio

It may be observed that the strength versus water-cement ratio curves for storage at 18°C (Fig. 3.11) is dissimilar from the usual curves for Portland cement concretes. This is characteristic of concretes made with HAC, and has been confirmed also for cylinders both of standard size[62] and other height-diameter ratios[8].

The values shown in Fig. 3.11 are for two cements only, and clearly some variation would be found with different cements, but the pattern of behaviour is the same in all cases. It is important to note that the residual strenth of mixes with moderate and high water-cement ratios, say over 0.5, may be so low as to be unacceptable for most structural purposes.

Tests[17] reported at the RILEM International Symposium on Concrete and Reinforced Concrete in Hot Countries in July 1960 showed that with a water-cement ratio of 0.65—a value not unknown in practice, certainly in those days—the strength of HAC concrete (a 1:6.2 mix) can be as low as 5.2 N/mm^2. A wider range of results is shown in Fig. 3.12, the conversion having been obtained by storage in water at 40°C for 100 days. The mixes shown in this figure were as follows:–

water-cement ratio	*aggregate-cement ratio*
0.65	6.2
0.45	4.0
0.35	3.0
0.29	2.0

Figure 3.12 **Relation between the strengths of different mixes of HAC concrete after 100 days at 18°C and at 40°C.**

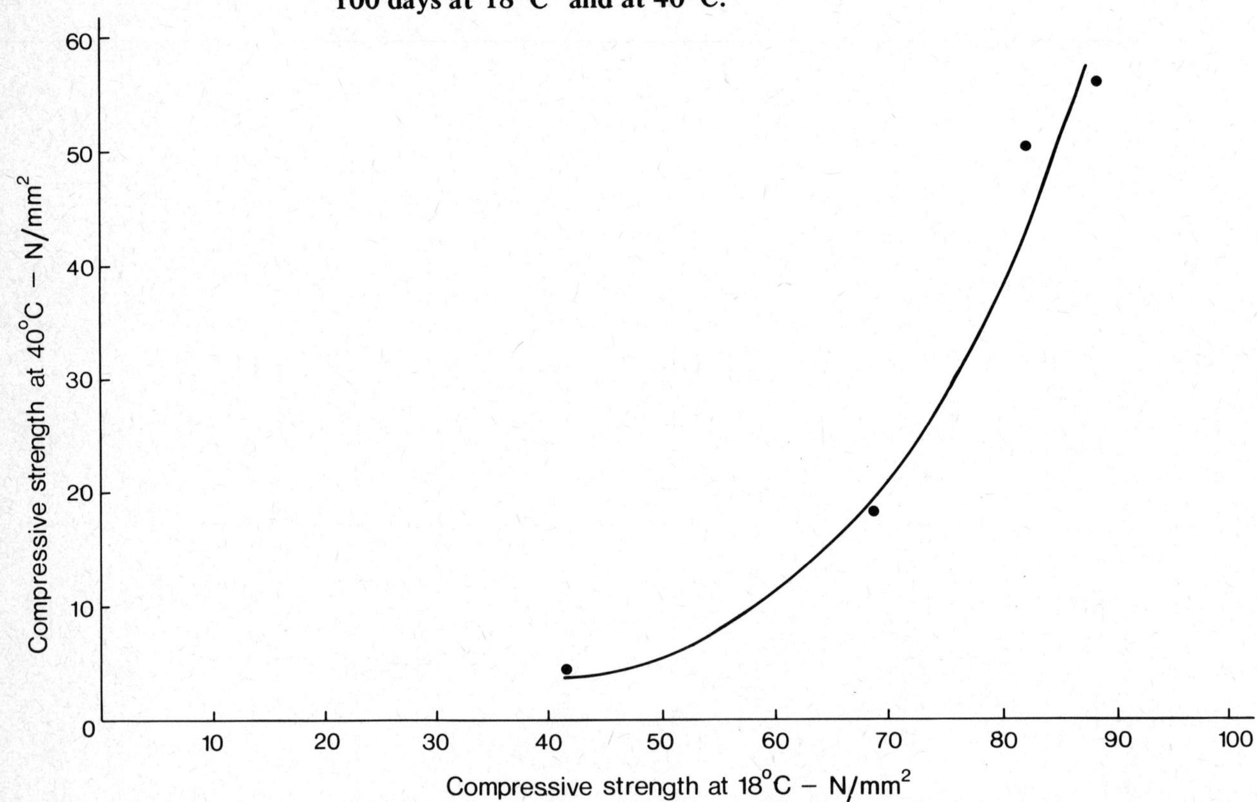

Not only the water-cement ratio but also the richness of the mix may affect the loss of strength on conversion. In 1958, we suggested[33] that for a given water-cement ratio, the leaner the mix the lower the porosity of the paste and, consequently, the lower should be the relative loss in strength with time for warm storage; our tests on mortar indicated that this is indeed the case.

In a 1960 paper[17] we pointed out that, while the strength of the very rich mixes with the extremely low water-cement ratios remained high, this was not so in the middle range; the mix with a water-cement ratio of 0.45 had a residual strength of 18.8 N/mm^2 and this would be inadequate in prestressed concrete although not necessarily dangerous in many other applications. Of course, in the majority of practical cases with which we are concerned HAC is used in prestressed concrete (see Section 2.5).

The practical significance of the rôle of the water-cement ratio in conversion is discussed in Chapter 5 but some laboratory data are of interest here. Tests[16] made by the French manufacturers of HAC show that at extremely low water-cement ratios the effects of conversion are small, or indeed the progress of conversion is limited. This is evident from Fig. 3.13 which shows the results for neat cement paste. The value of the water-cement ratio of 0.20 is, however, of no practical interest, and even 0.30 is of very limited application.

Figure 3.13

Relation between penetration resistance of HAC paste and water-cement ratio for different storage temperatures[16].

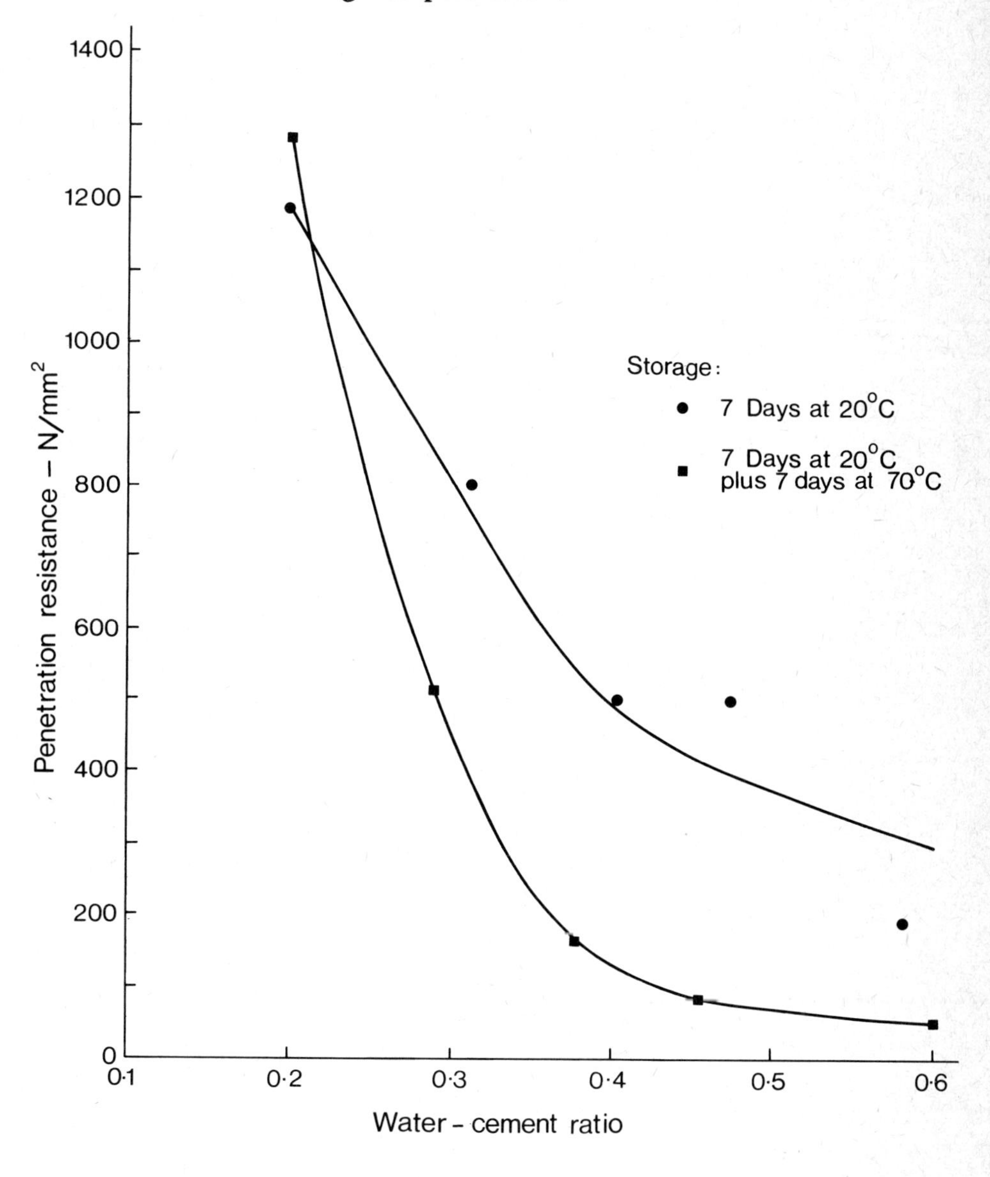

A different picture is given by some work done at the Building Research Establishment[14]. Test results on the strength of 13 mm cubes of neat HAC paste show that even at a water-cement ratio of 0.30, the strength at 50°C is only 34 per cent of the strength at 18°C; at a water-cement ratio of 0.50, the fraction is only 20 per cent. These values are much lower than those quoted by the manufacturers, in research sponsored by them[34], and indeed in the codes of practice, but we must stress once again that the data (reproduced in Fig. 3.14) refer to neat HAC pastes.

Figure 3.14 **Influence of the water-cement ratio on the compressive strength of fully converted cubes of neat HAC paste expressed as a percentage of strength at 18°C (13mm cubes)[14].**

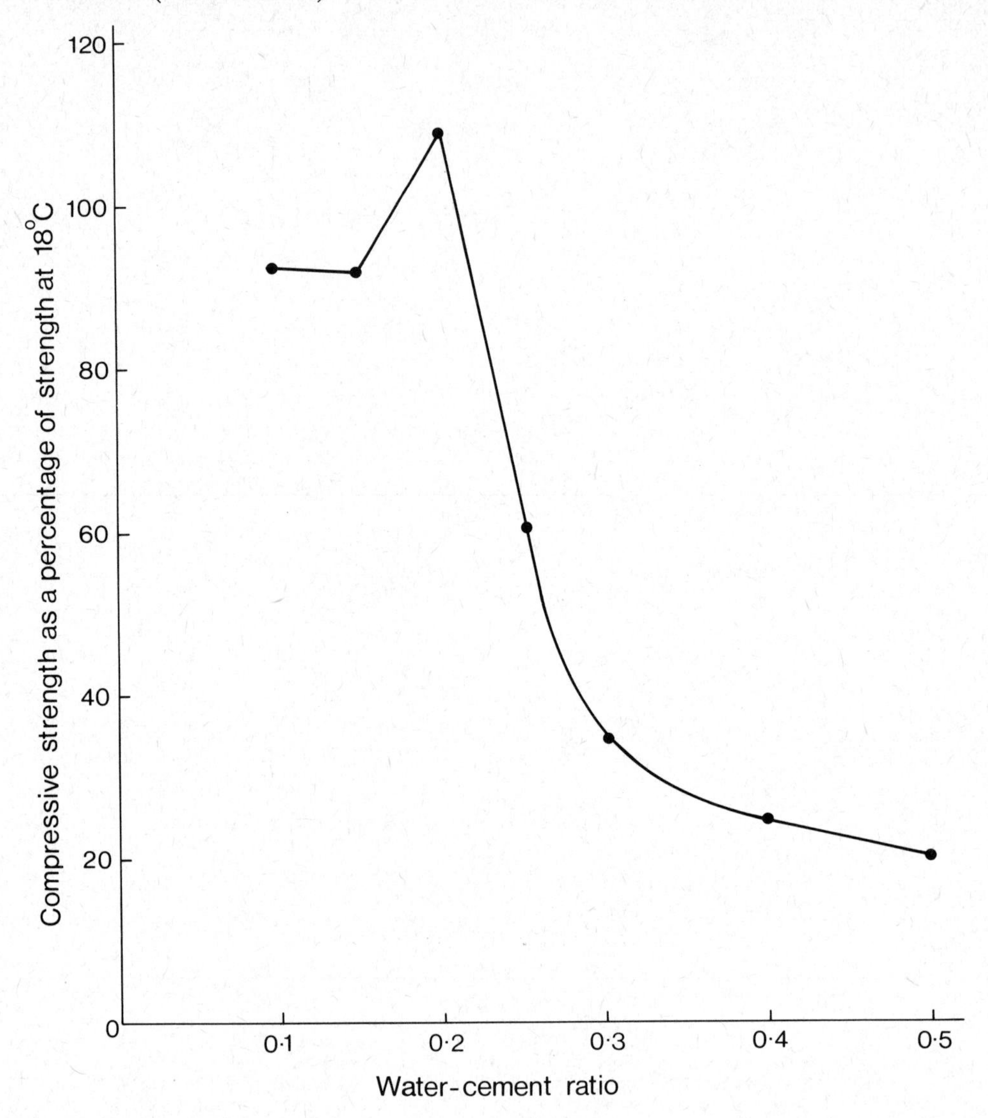

Minimum strength after conversion

The value of the minimum strength after conversion is still controversial. For many years, we thought that the manufacturers' data and those from the tests sponsored by them did not show the worst that can happen. Our view is confirmed by recent data from the Building Research Establishment[35]. Fig. 3.15 gives the relevant values, and Fig. 3.16 shows a plot of estimated fully converted strength as a percentage of one-day 'cool strength' for different water-cement ratios. It is possible that even the data of the Building Research Establishment are somewhat optimistic because we have heard of percentages down to 12 in some cases but it is uncertain how high the water-cement ratio was on those occasions. The most recent data [120] of the Building Research Establishment are shown in Fig. 3.17.

It is difficult to understand why the persistent denials of loss of strength on conversion under usual conditions and, above all, of a loss of strength continued.For instance, in the closing words of the discussion on Newman's paper [34] presented in November 1959, he said: " . . . *except for concrete definitely subjected to hot, wet conditions for long periods, the fully converted*

Figure 3.15 **Strength of HAC concrete as a function of water-cement ratio.**

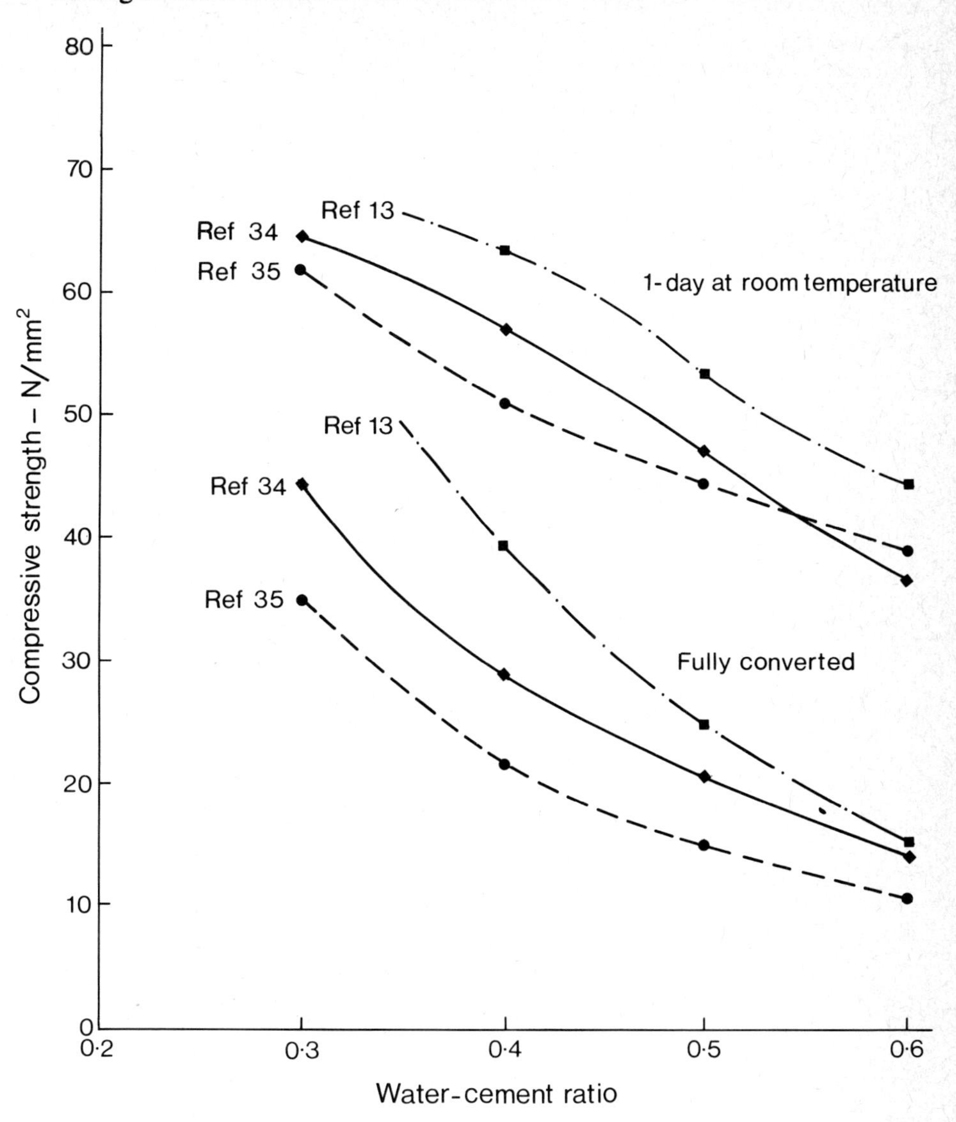

concrete was rare. As had been said previously, as long as the concrete was cured properly in the first 24 hours the conversion that was likely to occur was negligible."

Because in many publications (e.g. Ref. 36) it was stated that, provided early 'overheating' is avoided, there can be no later trouble due to 'warmth', our 1960 investigation was concerned with a delayed exposure to hot-wet conditions. The same mixes as mentioned on page 44 were used, and the specimens were stored in water at 18°C prior to exposure. Figs. 3.18 and 3.19 show that the residual strength was sensibly independent of the age at exposure to 40°C, which ranged between 6 hours and 84 days with intermediate values at 1, 3, 4, 7, 14, 28, and 56 days (only some of these are shown in the figures). Additional tests showed that even when the exposure is delayed for as long as six months the loss of strength is not abated and is complete in two months.

We performed similar tests at the temperature of 30°C; the rate of loss was lower, and considerably so in the case of concrete with a water-cement ratio of 0.45. It was, nevertheless, pointed out that [17] *"a continuous slow loss of*

Figure 3.16 **Influence of the water-cement ratio on the strength of converted HAC concrete expressed as a percentage of one-day strength.**

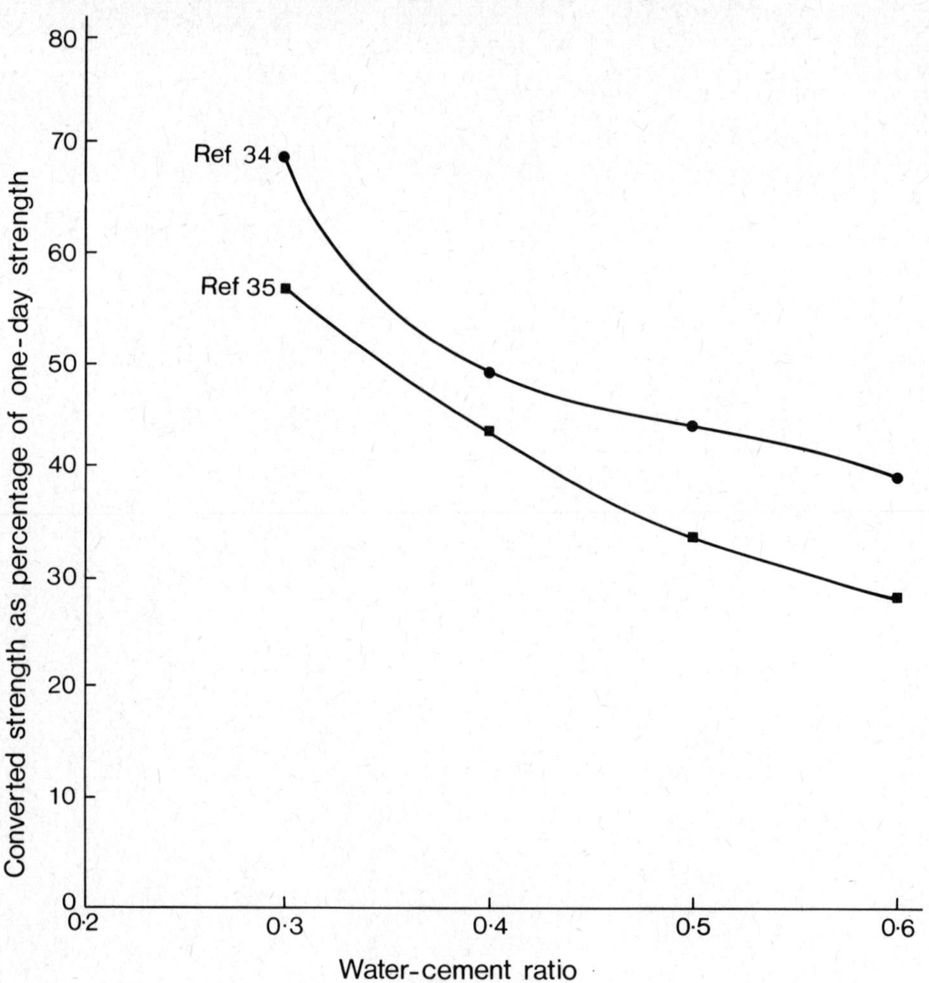

strength takes place, and after a sufficiently long time the strength will fall to the residual value of the same mix stored under hot-wet conditions from (the age of six hours)."

Tests on periodic exposure to a higher temperature, reported in the same paper[17], showed that the loss of strength is independent of the age at which the higher temperature acts and is cumulative so that HAC never ceases to be vulnerable to conversion. Exposure to hot and humid air by storing the specimens above water showed no significant differences in the loss of strength, and it was reported that whatever the age at which the hot-humid conditions begin to act, and whatever the combination of periods of exposure, a loss of strength right down to the residual value takes place. One cannot help thinking of the relation of this statement to the events in the Stepney swimming pool (see Chapter 7).

Figure 3.17 **Influence of water-cement ratio on the long-term strength of HAC concrete stored at 18°C and at 38°C[120].**

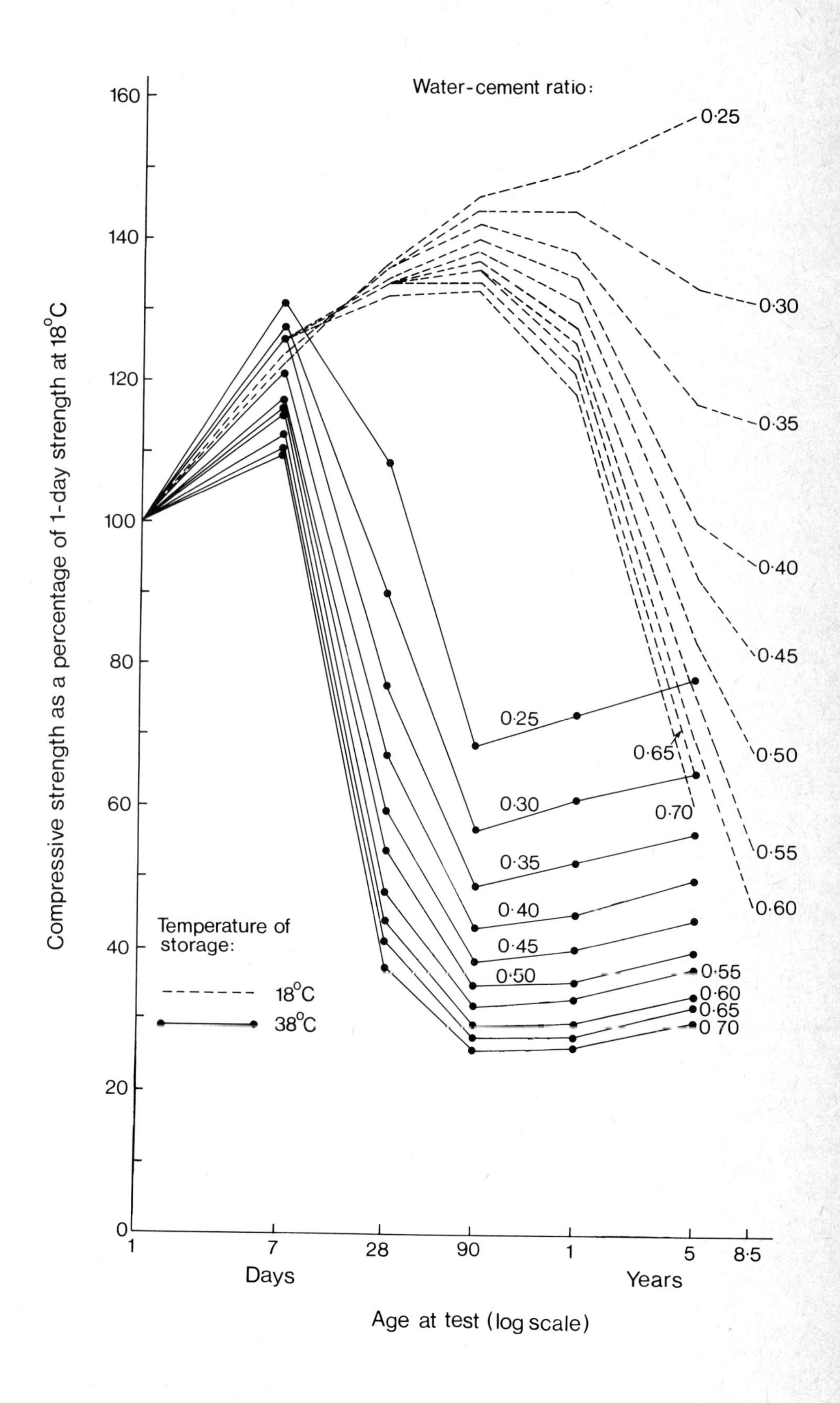

Figure 3.18 **Strength of 1:6.2 HAC concrete with a water-cement ratio of 0.65 transferred to water at 40°C at different ages.**

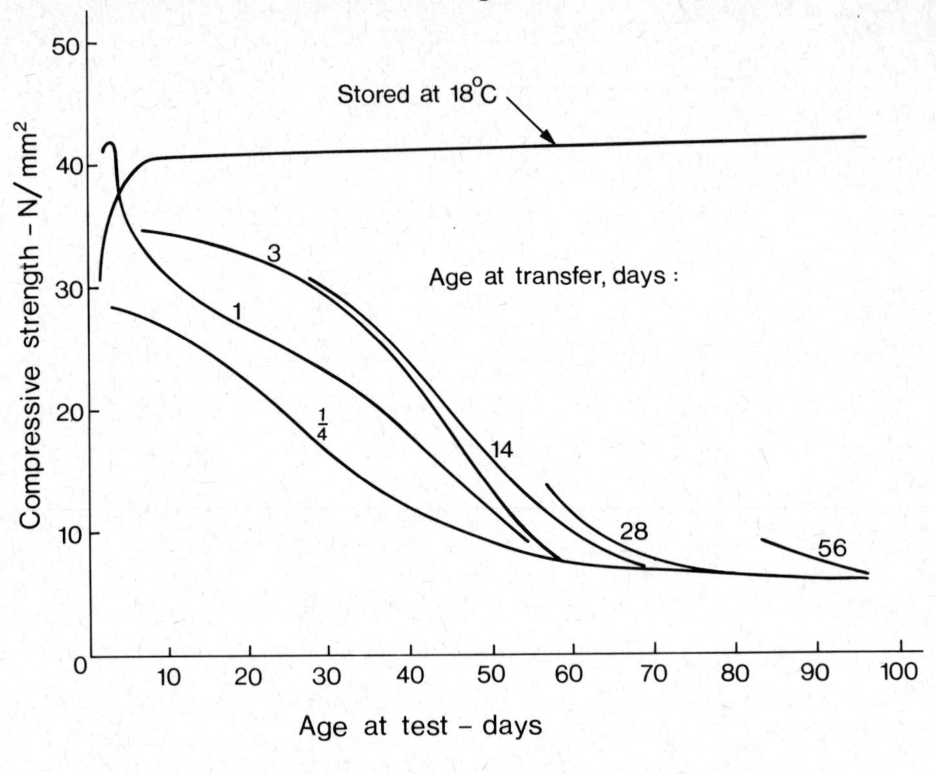

Figure 3.19 **Strength of 1:4 HAC concrete with a water-cement ratio of 0.45 transferred to water at 40°C at different ages.**

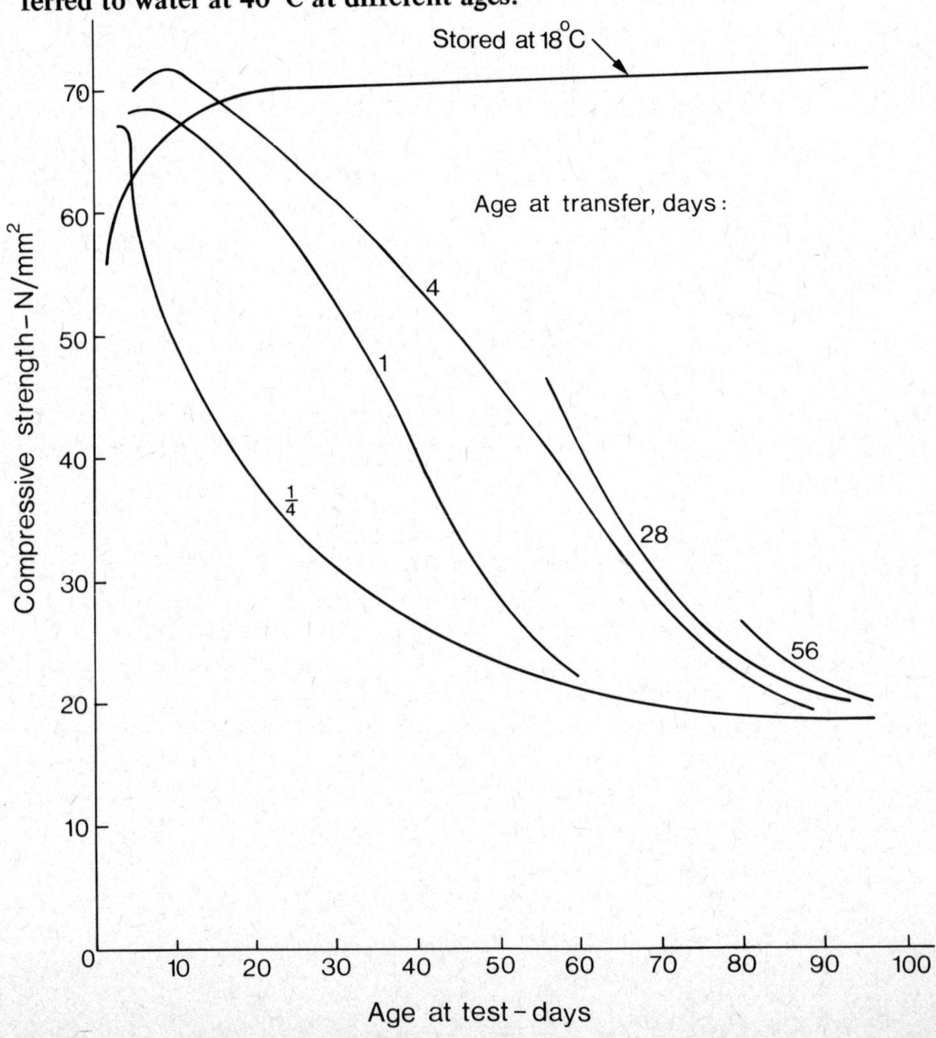

4. Long-term Laboratory Tests

4.1 Behaviour of test specimens

It is important to distinguish between problems arising in HAC due to conversion on 'overheating'soon after casting and the long-term changes. It is the latter, i.e. the slow or delayed conversion, that is considered in this chapter. We shall deal with laboratory tests as distinct from experience in actual structures, which forms the subject matter of Chapters 5 to 7.

Laboratory tests and behaviour in actual structures differ from one another very significantly insofar as direct and background information about the concrete are concerned. In the case of laboratory tests, the details of mixes and the conditions of exposure are known exactly; however, the exposure may not be representative of the conditions obtained in practice. On the other hand, a study of the evidence from actual construction cases enables us to see the exact extent of deterioration of concrete and of damage to structural members under practical exposure conditions. We are, however, uncertain about the mix used and about the conditions during the whole life of the structure; in particular, some doubt may exist about the exact water-cement ratio, about the workmanship, or about the efficiency of curing or other means of control of temperature at the time of casting. This is an important point as in some cases of deterioration of structures, even at an advanced age, some people blame the conversion at the time of casting or even workmanship in general rather than admit the existence of long-term changes. After a lapse of many years the early causes can be neither proved nor disproved, and this has given rise to the school of thought: 'HAC concrete never deteriorates – it is only sometimes badly made'. More about this is said in Chapter 5.

To discuss rationally this proposition it is useful to consider in detail the behaviour of HAC concrete test specimens under carefully controlled conditions as, from such test results, one can infer the probable effects of certain conditions in actual structures and the probable conditions which produced some observed effects.

Let us then look at some laboratory tests. We consider as fully accepted the view that the conversion of the hexagonal calcium aluminate hydrate to cubic form occurs not only at temperatures above some critical value but also at room temperature, i.e. in the region of 18 to 20°C (see Chapter 3). Since the original hydrate is chemically unstable, the change is spontaneous and irreversible, and cannot be prevented. The rate of conversion is lower the lower the temperature, and conversion is indeed very slow at ordinary temperatures. For this reason, a change in strength can be detected only by long-term tests.

From their long-term nature it follows that test results became available rather later than information on conversion at the time of casting. Some of the earliest tests were conducted at the Building Research Station, and their analysis was published in 1960 [37]. The specimens used were 102 by 203 mm cylinders. A wide range of mixes was investigated. The specimens were stored in water and in air outdoors in Hertfordshire for ten years.

A statistical analysis of the test results for specimens stored in water for ten years indicates that the strength of HAC concrete decreases significantly with time after the first five years. Storage in air shows in some cases a similar behaviour, and a slight loss of strength beyond the age of five to ten years was observed in specimens stored at a relative humidity as low as 65 per cent and a temperature of 18 °C. Although the loss appears small, if continued over a number of years, it would lead to an important reduction in strength. This is evident from Table 4.1, which is based on the results of the Building Research Station [37]. It can be seen that the average loss of strength on water storage is 2.1 per cent per annum; at this rate, the 25-year strength would be 58 per cent of the 5-year strength, and this is clearly significant from the standpoint of the strength of a structure and of its serviceability. We should point out that the actual values of strength quoted are rather low as the cements used were manufactured between 1925 and 1931, and the more modern cements are of higher strength.

Table 4.1 **Long-term strength of concrete specimens[37]**

Cement +	*Storage Condition*	*Water-cement ratio*	*Compressive strength at the age of 5 years* N/mm^2	*Average loss of strength per annum between the ages 5 and 20 years as a percentage of 5-year strength*	*20-year strength as a percentage of 3 to 12 months strength*
A		0.60	19.4	1.2	87
B	Air *	0.60	18.1	– 0.2	70
C		0.43	29.4	1.1	67
A		0.60	35.4	2.6	57
B	Water	0.60	27.9	2.6	69
C		0.60	31.2	1.1	97

\+ Cements manufactured between 1925 and 1931

* Specimens stored at 12 to 14°C during the first six years and thereafter at 18°C and a relative humidity of 65 per cent

A second set of data on the long-term behaviour of test specimens (76 mm by 152 mm cylinders) was published also in 1960. This was a report[38] of the Institution of Civil Engineers on an investigation of the durability of concrete piles in sea water. Two cements and five mixes were used with water-cement ratios ranging between 0.29 and 0.85. The specimens were moist-cured for seven days, then stored in air up to the age of 28 days, and thereafter in fresh water. The ten- and five-year strengths are plotted against one another in Fig. 4.1, and it can be seen that in all cases but one there was a decrease in strength between these two ages. The decrease is admittedly small but a statistical investigation suggests that in the majority of cases it is significant.

Figure 4.1 **Comparison of five-year and ten-year strengths of HAC concrete stored in fresh water.**

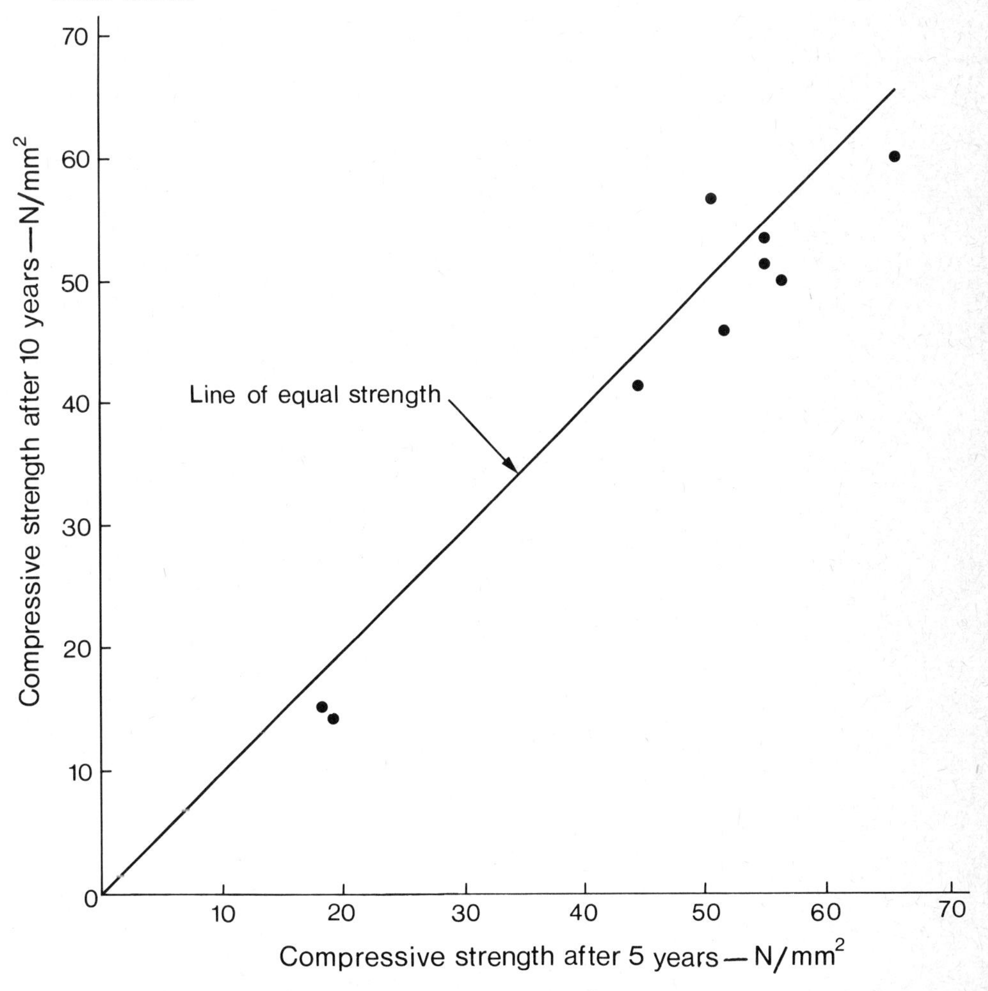

No other British long-term tests have been reported but there are available numerous foreign investigations. Admittedly, they deal with cements made abroad but to reject test results merely because they are foreign is not sensible. In the case of Portland cement, we use data from Europe and the United States as well as those obtained at home. As far as HAC is concerned, it was shown in Chapters 2 and 3 that cements of a wide range of composition and method of manufacture all behave broadly in the same manner: their products of hydration and the rate of development of strength are similar. The only significant difference to be noted is the presence of an appreciable quantity of sulphides in the German cement.

Figure 4.2

Long-term strength of HAC concrete stored in laboratory air and outdoors in Hungary.

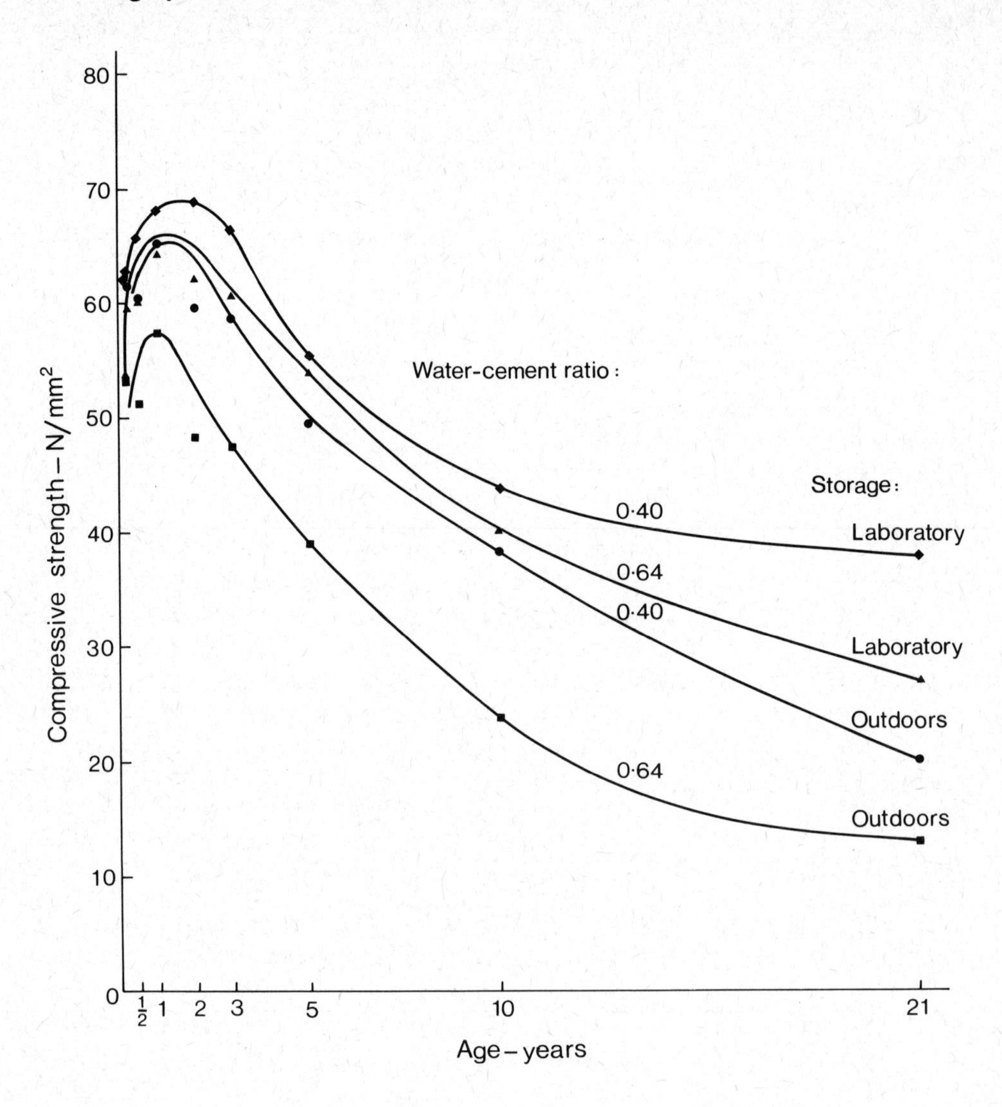

Table 4.2

Long-term strength of HAC concrete stored in air in Hungary[25]

Storage	*Water-cement ratio*	*Workability*	*Strength at 21 years N/mm²*	*Strength at 21 years as a percentage of maximum strength*
Laboratory	0.40	low	38.1	55
	0.64	high	26.8	41
Outdoors	0.40	low	20.0	31
	0.64	high	13.3	23

Some Hungarian data[25] are shown in Fig. 4.2. The specimens used were 200 mm cubes, stripped at 24 hours, cured under moist hessian for seven days, then stored in the laboratory at 18 to 20°C for 28 days and thereafter either continuing in the laboratory or stored in open air in Budapest. The cement content was 267 kg/m^3. Fig. 4.2 makes it clear that a large loss of strength occurred, especially in the concrete kept outdoors, probably because of a higher temperature and possibly also a higher humidity than indoors. A summary of the results is given in Table 4.2; this confirms that the loss is greater when the water-cement ratio is higher.

To establish the relevance of the Hungarian results to British conditions, a comparison of climatic data should be made. The average monthly temperature in Budapest is 21°C maximum and –2°C minimum, compared with 17°C and 4°C respectively in London. The corresponding values for rainfall are 660mm and 620mm respectively. It seems thus that the Hungarian conditions, although slightly more severe, are not very different from those in the United Kingdom.

The Hungarian data are complemented by a study of neat cement pastes made with HAC from five countries. The data are summarized in Table 4.3. The majority of pastes reached their maximum strength between the ages of six months and one year, and thereafter showed a gradual loss. In a neat cement paste, the water-cement ratio would have to be very low, probably below 0.3, in all cases, and consequently the loss of strength was small but by no means negligible.

The loss occurred in *all* the cements, including a white HAC (see Section 2.1), water storage in most cases leading to a somewhat greater loss than air storage. This agrees with our earlier observations.

The data in Table 4.3 show a considerable variation but it must be remembered that the cements tested were made at a number of works, using different raw materials and several methods of manufacture. It is therefore only to be expected that no *single* pattern in the strength—time changes is present but there is repeated evidence of the gradual decrease in strength over long periods.

Some of the tests discussed above were made on neat cement paste and on sand-cement mortar. It is important to be sure that the behaviour of HAC in *concrete* is no different, and we should therefore report more test results on concrete. Table 4.4 shows such data for 200 mm cubes with cement contents between 252 and 303 kg/m^3, moist cured for seven days and then stored outdoors. Cements made in several countries were used in these tests[25], and they all showed the same pattern of strength. It may be noted that in some cases a very low water-cement ratio was used (0.32, 0.38) and yet even there the minimum strength expressed as a percentage of maximum strength was 46, 53 and 54. This should tend to disprove the argument that, provided we use a low water-cement ratio, HAC is totally satisfactory so that the 'HAC problem' can be avoided by the use of a good quality concrete and good quality control (see Chapter 10).

In another series of tests[39], 100 mm cubes of a 1:5.4 concrete made with a German HAC and a water-cement ratio of 0.55 were stored in air at 20°C and a relative humidity of 65 per cent after 7 days' moist curing at 20°C. The pattern of change in strength was as follows:

Table 4.3 **Long-term strength of neat cement pastes made with different foreign HACs[25]**

Country of manufacture	*Water-stored ∮*			*Air-stored*		
	Minimum strength N/mm²	*Age at which observed years*	*Minimum strength as a percentage of maximum strength*	*Minimum strength N/mm²*	*Age at which observed years*	*Minimum strength as a percentage of maximum strength*
France	25.0	2	41	46.5 ∫	5	57∫
France	26.8	5 *	34	42.0	2	59
France	25.5	5 *	32	48.1 ∫	5 *	54 ∫
France	90.6	2 *	90	79.0	2 *	80
Switzerland	33.0	5	41	47.7 ∫	2	57 ∫
Germany	38.6	2	43	76.0	5	71
Germany	54.2	5 *	57	68.9	2	86
Germany	36.4	5 *	46	49.7	5 *	55
Germany	67.2	5 *	73	48.8	2	57
Czechoslovakia	65.3	5 *	57			
Czechoslovakia	98.5	5 *	95	64.5	5 *	63
Hungary	85.2	10 *	90	58.5	10 *	67
Hungary †	90.6	5 *	84			
MEAN			60			64

* Maximum age at which strength determined ∫ Mixed storage ∮ at 18-20°C † White HAC

Country of manufacture	Cement	Water-cement ratio	Compressive strength (N/mm^2) at 7 days	28 days	1 year	3 years	5 years	7 years	10 years	20 years	24 years	Minimum strength as a percentage of maximum strength †
Hungary	A	0.52	61.0	65.6	65.6	60.1 *			31.9	33.5	28.1	40
France	B	0.52	61.9	71.2	71.2	41.1 *			31.6	34.1		41
France	B	0.52			66.1	34.2 *			32.0	29.0		44
Germany	C	0.51		69.4	73.4	68.1	48.1		30.4			41
Germany	C	0.58		60.0	69.0	66.6	46.2		30.4			44
Germany	D	0.38	51.6	59.8	68.9	63.6	49.9	43.4	40.6	36.5		53
Germany	D	0.48	44.9	52.1	56.3	47.1	39.9	38.4	37.5	33.0		59
Germany	D	0.56	51.8	57.5	71.1	66.9	55.0	37.9	32.4	28.0		39
Germany	D	0.62	46.2	53.7	60.0	56.8	47.1	42.5	32.9	29.9		50
France	E	0.46	70.2	79.6	74.0	78.6	79.6		69.4		49.1	62
France	E	0.55	60.0	66.6	72.1	75.8	70.2		59.2		41.0	54
Czechoslovakia	F	0.32	64.3	70.2	75.8	55.9	40.6	43.4				54
Czechoslovakia	F	0.38	55.4	63.4	70.2	54.1	39.2	32.3				46
Czechoslovakia	F	0.48	75.8	86.0	91.4	60.5	33.7	30.4				33
Czechoslovakia	F	0.56	60.7	73.0	79.6	72.1	36.5	28.6				36

* Strength measured at 4 years

† Note that the maximum strength may be higher than at the age listed in the table

Table 4.4 **Strength of concrete made with different HACs and stored outdoors in Hungary[25]**

Age	3 days	28 days	3 months	3 years
Strength N/mm^2	60.6	80.5	87.4	49.9

Our evidence on the pattern of behaviour of HAC should cover as many cements and exposure conditions as possible. It is therefore worthwhile to look at the results of tests on 1:3 concrete cylinders (51 mm by 102 mm) made with French and American cements and stored in tap water and in lake water[40]. Table 4.5 shows that for concrete with a water-cement ratio of about 0.52, the average 25-year strength was 88 per cent of the seven-day strength. On the other hand, storage in Medicine Lake in South Dakota, which has an average salt content of 4.79 per cent (largely in the form of magnesium and sodium sulphates), resulted in little or no loss of strength. Our explanation is that in all likelihood the lake temperature was considerably lower than the tap water temperature.

An important point which arises from the data of Table 4.5 is the extremely good resistance of HAC to sulphates. This is of course well-known and is discussed in Section 2.2 but it may be of interest to include here another set of data [25] indicating that HAC concrete withstands better exposure to a cool sulphate solution than to tap water at room temperature. Fig. 4.3 shows these results together with those for concretes stored in the

Figure 4.3 **Long-term strength of HAC concrete stored under different conditions.**

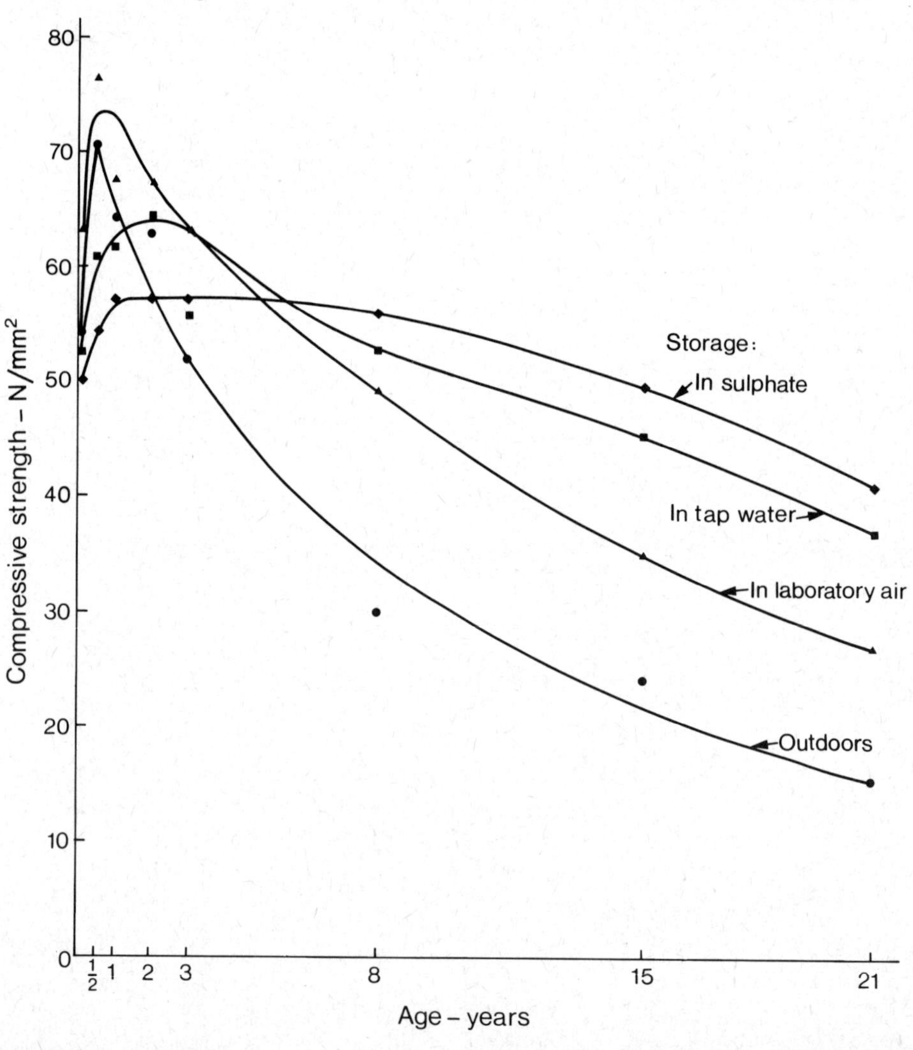

Country of manufacture	Cement	Water-cement ratio	Concrete stored in tap water						Concrete stored in lake			
			Compressive strength (N/mm^2) at									
			7 days	28 days	1 year	5 years	10 years	24-25 years	1 year	5 years	10 years	20 years
France	A		24.1	32.4	26.1	26.3						
France	A	0.62	32.8	33.5	37.6	23.3			38.8	44.6		42.2
United States	B	0.53	53.3	49.9	51.6	27.8		39.2	48.3	50.6		36.1
United States	C	0.51	46.2	46.8	43.2	53.6		48.1	51.1	61.7		59.1
United States	D		49.2	47.0	53.7	53.4		46.3				
United States	D	0.53	48.8	45.7	56.3	47.5		41.1	56.6	54.0		47.5
United States	E	0.53	45.9	49.1	55.9	48.6			52.3	53.2		61.4
United States	E	0.59	47.5	49.7	59.5	43.3			51.0	46.8		55.5
United States	E	0.66	48.8	53.7	53.9	44.4			49.8	53.9		49.7
United States	F	0.62	44.1	49.0	37.9	36.1	39.6		44.1	45.9	42.2	
United States	G	0.62	45.1	45.2	31.7	39.3			35.2	43.2	52.2	
United States	H	0.62	37.3	48.1	43.3	32.6			43.0	48.9	36.8	
MEAN			43.6	45.8	45.9	39.7			47.0	50.3		

Table 4.5 **Strength of concrete made with different HACs and stored in water**[40]

laboratory and outdoors. In all the tests, the specimens used were 200 mm cubes, the concrete having a water-cement ratio of 0.60 and a cement content of 250 kg/m^3. From Fig. 4.3 it is apparent that the outdoor air exposure led to the highest loss of strength, with laboratory air second in the extent of loss. The explanation probably lies in the fact that the temperature was higher outdoors than in water storage. Since a significant loss took place in all cases, it seems that the moisture conditions were adequate for the necessary reactions; the outdoor exposure was in Budapest where the mean of the average monthly relative humidities is 68 per cent; the comparable figure for London is 72 per cent.

High water-cement ratios are now discredited but for the sake of completeness some test results on HAC concrete with a water-cement ratio of 1.13 may be of interest [25]. Such concrete stored in sulphate-bearing water suffered, even cool, damage (see Table 4.6),probably because of its more porous structure. This likely influence of initial porosity, and of an increase in porosity due to conversion, on durability is of interest and is discussed more fully in Section 4.7. Here we may mention that experiments[14] on neat HAC paste (13 mm cubes) showed that conversion alone is not a measure of liability to chemical attack but, if conversion takes place so that porosity of the concrete increases, the two are closely related.

Table 4.6

Loss of strength of HAC concrete with a water-cement ratio of 1.13[25]

Storage condition	*Strength at the age of 21 years N/mm^2*	*21 year strength as a percentage of maximum strength*
Laboratory air	12.8	42
Outdoors in Budapest	7.0	31
In sulphate-bearing ground	15.1	73
In tap water	9.4	41

Few British test results on the long-term strength of concrete are available but some time ago we made a 102 mm by 102 mm by 406 mm prism of 1:6 concrete with a water-cement ratio of 0.57 and stored it in the very soft Manchester water at a temperature never above 20°C. The prism deteriorated after four years to such an extent that the concrete could be picked out by hand to a depth of 6mm, and more at the corners. The interior of the prism had a compressive strength of 19.2 N/mm^2, compared with 53.1 N/mm^2 at 14 days. X-ray and differential thermal analyses showed that high conversion had occurred.

One more series of long-term tests deserves mention as the test specimens were made on the site of an actual construction but were subsequently stored in the laboratory[18]. Yugoslav cement was used with a water-cement ratio of 0.60 and cement contents of 400 and 300 kg/m^3. The concrete was cast in November 1949 during the construction of a bridge on the Belgrade-Zagreb railway, when the air temperature ranged from 6°C to 14°C. The specimens were wet-cured for seven days and then stored on site (in winter) for two months. Thereafter, they were stored in the laboratory at 17±2°C and a relative humidity usually of 65±5 per cent. The loss of strength in the first three to four years was followed by a recovery of strength; this is discussed in

Section 4.2, and Fig. 4.5 shows the shape of the strength—time curve for the richer mix. The leaner mix behaved in a similar manner but the strengths were somewhat lower. Since the water-cement ratio was the same in both mixes, the richness of the mix appears to be the factor as in the case of our tests[33], discussed in Section 3.3.

All the preceding test data refer to the compressive strength of concrete and of cement paste. Tests on the tensile strength of neat cement pastes (made with the cements listed in Table 4.3) have shown that the loss of strength is less regular. Tests[41] on mortar made with a French-produced HAC and a water-cement ratio of 0.65 suggest that the losses of strength in tension and in compression are comparable (see Fig. 4.4).

Figure 4.4 **Compressive and tensile strengths of HAC mortar stored in water at 20°C.**

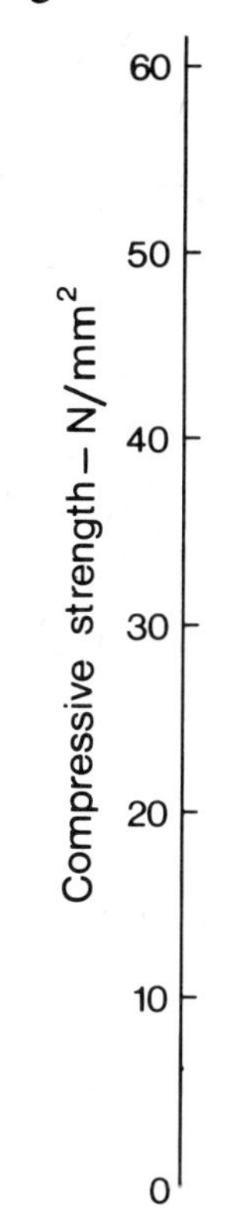

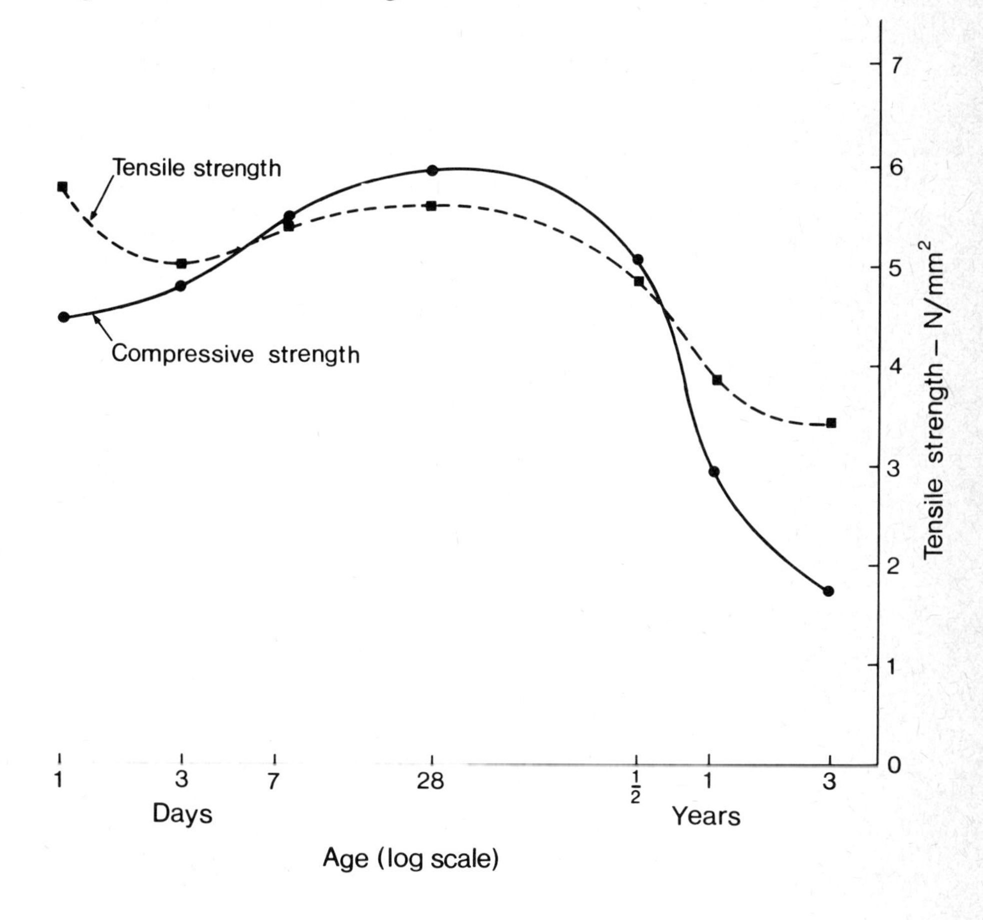

4.2 Recovery of strength

In the discussion of the loss of strength associated with conversion it was pointed out on a number of occasions that the decrease in strength is followed by its gradual recovery. Before commenting on the practical significance of such a pattern of strength—time development, let us present some actual data.

In the Yugoslav tests[18] referred to in the preceding section, concrete specimens made at 6 °C to 14 °C outdoors were stored in the laboratory at

17±2°C and a relative humidity of 65±5 per cent. The pattern of strength development over ten years is shown in Fig. 4.5 where each point represents the average value for 21 sawn cubes made from the mix with a cement content of 400 kg/m^3. The recovery of strength is unmistakable. The shape of the corresponding curve for a mix with a cement content of 300kg/m^3 is similar but somewhat lower strengths were recorded. Both mixes had the same water-cement ratio of 0.60 so that we are dealing with the effect of richness of the mix.

A small recovery of strength has been observed also in some other tests. This was, for instance, the case with several of the neat cement paste specimens tested in Hungary[25], but the great majority showed no recovery. An explanation of this is offered later in this section.

British tests[34] on HAC concrete cubes (102 mm) cured for six hours at room temperature and then stored in water at 38°C showed an increase in strength of about 4.0 N/mm^2 for mixes with water-cement ratios of 0.35, 0.45 and 0.55 and corresponding aggregate-cement ratios of 3, 4.5 and 6; there was virtually no increase in the case of the 1:7.5 mix with a water-

Figure 4.5 **Long-term strength of HAC concrete stored in the laboratory in Yugoslavia.**

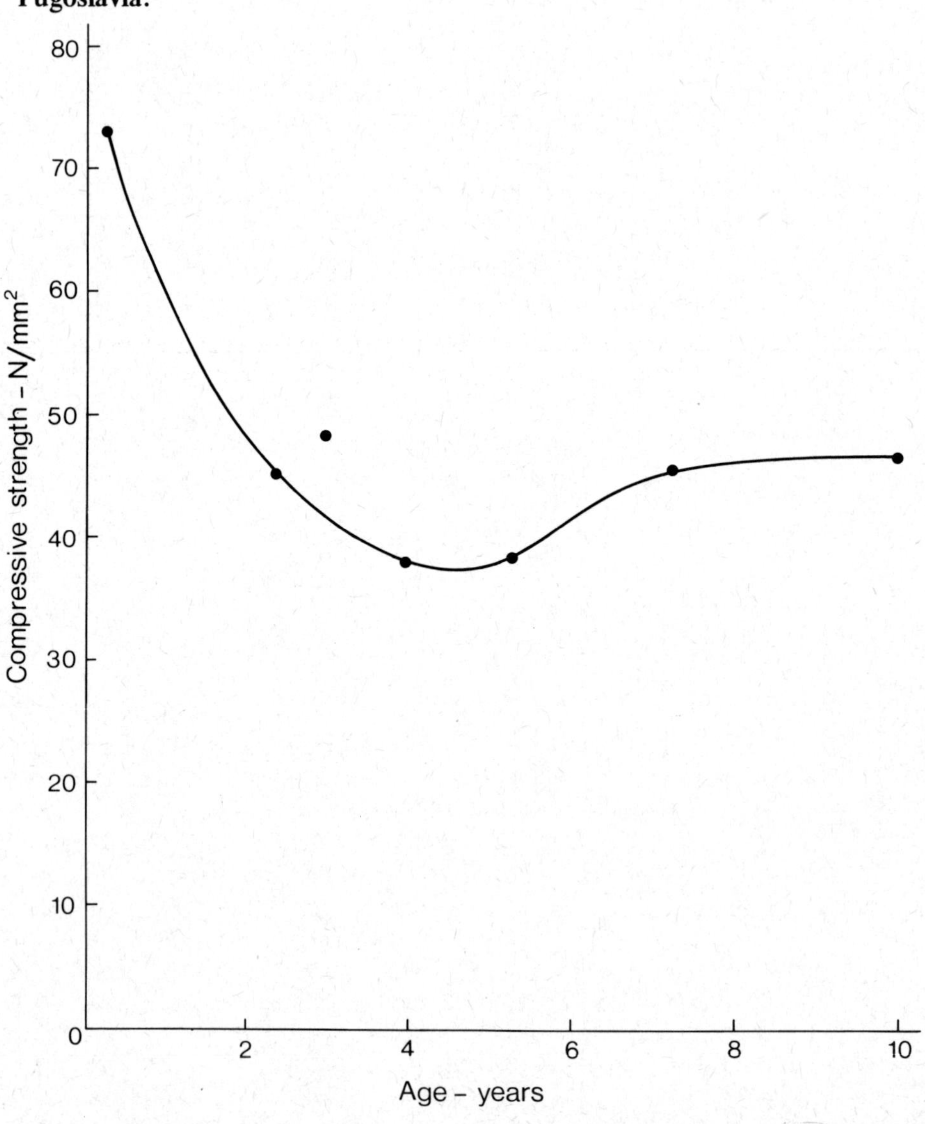

cement ratio of 0.65. When the temperature of 38°C was applied from the time of casting there was an increase of about 3.0 N/mm^2 in the case of the 1:3, 0.35 mix, 1.8 N/mm^2 for the 1:4.5, 0.45 mix but virtually no increase at the higher water-cement ratio.

An increase in strength following conversion was indirectly demonstrated by data[42] on the dynamic modulus of elasticity. Conversion was induced by storage in water at 38°C. There was a later increase in the modulus for the 1:2:4 mix with a water-cement ratio of 0.35. However, when the water-cement ratio was 0.55, the increase was negligible. Fig. 3.17 shows some more recent data on the recovery of strength with time.

It seems thus that when rapid conversion occurs, a large proportion of the cement remains unhydrated. Further gradual hydration then takes place and this produces new hydrates which occupy some of the voids induced by hydration. As a result, a denser hydrated cement paste is obtained and this naturally has a higher strength than at the time of completion (or nearly so) of conversion. We confirmed this by boiling concrete with a water-cement ratio of 0.45 at the age of 24 hours and then storing it in cold water. There was a recovery of strength of about 3.4 N/mm^2 during the next few days and in the following eight months there was a further gain of up to 1.7 N/mm^2 in some cases.

We should stress that recovery of strength is possible only if conversion takes place when a considerable amount of unhydrated cement is still present: further hydration would fill some of the pores caused by conversion and some rise in strength would result. This is probably more likely to take place in rich mixes with low water-cement ratios, where a greater amount of unhydrated cement is present at any age, and it is indeed in this type of mix that an apparent recovery has been observed.

Further hydration requires the presence of water. From considerations of the amount of water that can become chemically fixed, it appears that at a water-cement ratio between 0.3 and 0.4, no free water will be left[16]. Theoretically, this would mean achieving hydration to a point when further conversion does not increase the porosity of the hydrated paste, but the practical significance of this is doubtful because there may be an intermediate period of higher porosity and lower strength. A great deal will also depend on the possibility of ingress of water during the life of the concrete. In the case of neat cement paste (see earlier), the permeability may be too low to allow the further ingress of water necessary for continued hydration. Anyway, neat cement paste is of no practical interest.

Conversion of well-hydrated cement is unlikely to be followed by any changes that may cause an increase in strength, and there is no proof that a reliably steady and significant build-up of strength follows the loss on conversion. In any case, even if recovery of strength took place in the form of a gradual increase in the strength of the converted cement paste, such a recovery would be of no practical consequence as far as structural concrete is concerned, as even a limited period in the life of a structure when concrete has inadequate strength is unacceptable.

Against this background, we can mention a rather curious article in the *New Civil Engineer,* published on 21st March, 1974, that is at the time of the mounting interest in the causes of failure of the Stepney school[43]. The author is anonymous and is identified merely as a former technical manager of

Lafarge Aluminous Cement Co. Ltd. The title is very fashionable: *No more wetness about warmth*. Expectations are aroused on reading the introductory note, which says that the *New Civil Engineer "gives an account of how the phenomenon of conversion may be exploited to produce durable high-alumina cement concrete."*

The paper suggests that the reaction of conversion (see Section 3.1) may create fissures in the previously impermeable mass of hydrates surrounding each residual grain of cement. These fissures could make it possible for water, which is generated by the reaction of conversion, to bring about a secondary reaction of hydration. The form suggested is

$$3[CaO.Al_2O_3] + 30H_2O \rightarrow 3CaO.Al_2O_3.10H_2O$$

which by the reaction of conversion changes to

$$3CaO.Al_2O_3.6H_2O + 2[Al_2O_3.3H_2O] + 18H_2O$$

Because of the nett reduction in the quantity of water, the author[43] suggests that *"there might be an overall reaction in which the otherwise deleterious water (helps) to fill the otherwise detrimental voids with useful solid hydrates"*. These changes are written as

$$3[CaO.Al_2O_3] + 12H_2O \rightarrow 3CaO.Al_2O_3.6H_2O + 2[Al_2O_3.3H_2O].$$

By considering the volume of the secondary products of hydration, the author[43] claims to establish *"the case where the amount of unhydrated cement is enough ultimately to produce the precise quantity of additional hydrates needed to make good the nett loss in hydrate volume caused by complete conversion of the original hexagonal hydrates."* He then disagrees with some work done at the Building Research Establishment[14] and assumes that $Al_2O_3.3H_2O$ has cementing properties. Hence he concludes that *"Any report of 'full conversion' suggests that the minimum strength point has long been passed."*

This *can* be the case though *need not* always be so as conversion may proceed even after all the cement which can be hydrated has been used up. Nevertheless, one can say that concrete which has undergone full conversion is unlikely to lose strength further although it can deteriorate through chemical attack.

The author of the paper[43] admits it is *"likely that storing mature concretes at successively increased temperatures would progressively steepen the hydrate-decay curve more than that of hydrate growth. Voids would for a time be created faster than they could be filled and transient minima in useful properties would occur earlier, and be lower, for each increase in storage temperature."* These transient and low minima of strength are surely unacceptable to the structural engineer.

On the basis of the above argument, the anonymous author[43] suggests hot curing to develop stable hydrates and states that this is in commercial use. This is a *volte face*: instead of preventing *overheating*, we should now use it deliberately to get over conversion and its ill-effects.

As a further corollary (*"formally unconfirmed"*) from his review, the author claims that he can calibrate the water-cement ratio for any specified minimum strength. This leads him to predict that the beams at the Stepney school *"would have withstood permanent service at 30°C. Unless the temperature can be shown to have exceeded this value a failure due to conversion seems inadmissible. A beam of similar quality in any case has been tested at the half-conversion point after storage at 45°C and found to bear load almost as well as a control beam stored normally."* It is very difficult to reconcile these claims with the findings of the report of the Building Research Establishment[35] on the Stepney school collapse (see Chapter 7), and we must be very wary of accepting the findings of this paper and believing that *"conversion may be exploited to produce durable high-alumina cement concrete"*. To an engineer, a cement that is not subject to conversion or to any other spontaneous deleterious changes is a safer material to use.

4.3 Alkali attack of HAC concrete

A detailed study of the chemical attack of HAC concrete is a matter for more chemically-oriented works but one form of attack deserves mention as it may be related to the failures described: this is the alkaline hydrolysis of calcium aluminates. This is as follows[24]:

$$K_2CO_3 + CaO.Al_2O_3.aq \rightarrow CaCO_3 + K_2O.Al_2O_3$$

CO_2 in the atmosphere regenerates the K CO :

$$CO_2 + K_2O.Al_2O_3 + aq \rightarrow K_2CO_3 + Al_2O_3.3H_2O$$

so that it can react again. The attack by the CO_2 thus continues, the alkali serving only as a carrier. The final reaction can thus be written as:

$$CO_2 + CaO.Al_2O_3.aq \rightarrow CaCO_3 + Al_2O_3.3H_2O$$

This form of deterioration can take place even in good quality concrete when the aggregate contains soluble alkalis as may be the case with granite[4], schist, and mica, especially when the aggregate contains crusher dust that may liberate the alkalis[24]. Under such circumstances, the deterioration may not be apparent from a simple exterior examination[44]. Loss of strength caused by alkalis derived from the aggregate and also through inadvertent inclusion has been observed in some structures in Switzerland. The French rules on aggregate, referred to in Section 8.1, are of interest in this context.

The alkali carbonate may also have its source outside the concrete, e.g. when this is placed in granitic ground or in Portland cement[44]. Deterioration occurs, of course, only if the HAC concrete is porous: the alkali carbonate then travels through the concrete and rises by capillary action to the surface higher up where concentration by evaporation and recarbonation takes place[24]. Dark stains on the surface can often be observed, sometimes followed by scaling, and the concrete in the interior becomes blackish in colour and soft[44].

A well-compacted and impermeable concrete would not be subject to this type of attack but conversion of calcium aluminate hydrate increases the porosity of concrete: thus even good quality concrete and one with an adequate residual strength would be liable to external alkali attack with a consequent serious deterioration. We consider, therefore, as unsound the argument that, if we design structures on the basis of the strength of fully converted concrete, there is nothing to worry about. Perhaps we should add that, when concrete is fully immersed, the movement of water through it would be small, so that the risk of damage is considerably smaller than immediately above water or in the zone between the tides, or under other conditions of alternating exposure to dry and humid atmosphere.

An example of alkali attack originating from Portland cement is offered by the HAC concrete poles supporting trolleybus wires in Moutiers in Savoy, France. These were erected between 1928 and 1930 and were embedded in Portland cement concrete. They deteriorated after several months, the damage occurring at the ground level and above up to a height of 50 mm to 400 mm: dark stains formed, followed by surface cracking and scaling, which increased in depth with time[4]. An investigation showed that the Portland cement had a high alkali content (by comparison, Portland cement manufactured in the United Kingdom usually has a low alkali content) and it was this that led to the reactions described earlier.

One further observation[4] is of interest: the presence of alkalis encourages conversion even in the absence of evaporation or carbonation.

The Stepney collapse

All this was reported in 1963[45], and was discussed by the designer and precast concrete manufacturer involved in the Stepney school[66], which was built during the years 1965 and 1966. According to the report of the Building Research Establishment[35], *"Feldspars and micas were detected in both the coarse and fine aggregate used for the concrete in the prestressed beams, and there was evidence that the high-alumina cement had hydrated in the presence of sodium and potassium ions which can be freed from these minerals by the alkali in the cement. The amount of 'freeable sodium and potassium' in the aggregate was 0.021 per cent, which is considered to be sufficient to increase markedly the rate of conversion and the associated loss of strength of the concrete."*

These findings are of interest as in a series of articles and notes on HAC in the *New Civil Engineer* there was one, under the title *Alkaline hydrolysis—the likely culprit*[47], which said *"It is beginning to look as though the prime culprit in last week's roof collapse at Stepney was the comparatively rare phenomenon of alkaline hydrolysis. The old bogey of conversion of high alumina cement can probably be ruled out . . . "* It is further said that the *"Key to the attack of the Pierhead joists at Stepney could well be the woodwool slabs which spanned between joists"* and *"Woodwool, made from wood shavings and Portland cement, usually contains as additives both calcium chloride and an alkali."* Hence it is suggested that *"the combination of alkali from both screed and woodwool could well have caused the trouble".*

A few weeks later, in an anonymous article in the *New Civil Engineer*[43], it was said with reference to the Stepney collapse that the presence of sodium and potassium hydroxides *"has been said to accelerate conversion, but it is possible that this is a case of the weakening effect of alkaline hydrolysis being wrongly ascribed to that of conversion (at an excessively high water-cement ratio)."*

So, it is not conversion, not HAC, but something else that causes 'trouble'—surely an understatement for a collapse of roof beams. And yet the report[35] of the Building Research Establishment says that the possibility of chloride attack was investigated and that *"the amount of chloride present in the ettringite was not, however, significant. The amount of chloride in the woodwool slabs in the screed and in the beams was also not significant. No signs were found of alkali hydrolysis, i.e. attack by alkali from the Portland cement in the screed."*

The report[35] states that sulpho-ettringite was present as a result of sulphate attack, the sulphate being derived from the gypsum plaster applied to the under surface of the woodwool slabs in the swimming pool. On the other hand, in the Stepney school gymnasium, *"no ettringite was found although some gypsum plaster had been used to make good the edges of the beams"*.

The actual rôle of chemical attack is assessed in the conclusions of the report[35] : *"the cause of failure . . . was loss of strength due to conversion of high alumina cement concrete followed by chemical attack leading to disruption of the concrete. The chemical attack, which was localised, was probably due to the action of sulphate, derived from the gypsum plaster in contact with the beams, taking place in the presence of water from condensation or leakage at the rooflights."* It is also stated[35] that *"the aggregate used which contained the minerals feldspar and mica had an accelerating effect on the rate of conversion and loss of strength."*

It seems thus that alkali hydrolysis is a rare cause of deterioration of HAC concrete, and can be fairly easily avoided: we must not allow percolation into HAC concrete through Portland cement and, above all, we must not use alkali-active aggregates.

5. Long-Term Failures of High-Alumina Cement Concrete Structures

5.1. Evaluation of data from structures

From Chapter 4 it is clear that HAC concrete, as tested in the laboratory, undergoes a serious loss of strength if at any time during its life it is exposed for a period to a temperature in excess of about 20°C to 25°C. It is also apparent that even below this temperature conversion at a very slow rate takes place, and there is strong evidence that storage outdoors under the conditions existing in many parts of Western or Central Europe leads to a gradual but marked, and sometimes serious, loss of strength. The very slow development of this loss should be noted. In 1963[45], we expressed the opinion that the loss may not become apparent until the age of five years, but within the next 15 to 20 years the loss is large; its actual value depends on the water-cement ratio of the mix and on the conditions of exposure. It was also stressed[45], that none of the long-term tests had failed to show a loss of strength.

Structural failures

These data covered a wide range of HAC concretes so far as the origin of the cement, composition of the concrete, and details of test conditions are concerned. Such data cannot be easily refuted but, not surprisingly, many engineers consider the behaviour of actual structures as the only fully convincing test of a material and a proof of its performance. After all, the size of the structural members is much larger than that of the specimens, and workmanship on site enters the picture.

Thus, ideally, the proof of the long-term loss of strength of HAC should come from actual structures, but we encounter two difficulties. First, when failure of a structure occurs, there is usually a not unnatural reluctance on the part of all concerned to publish and publicise the event. Moreover, financial responsibility is usually in dispute so that statements on causes of failure and therefore on blame are avoided. Confidential information can sometimes be obtained but must not be published.

The second difficulty is more complex. When failure has actually taken place and the deterioration of HAC concrete with age is suspected to be the cause, there are always raised counter-accusations, for instance, of poor workmanship or of improper curing and an early rise in temperature of the concrete. Thus, it is claimed that conversion occurred soon after casting and that with care all deterioration could have been avoided. This, of course, is difficult to disprove (or to prove) after a lapse of twenty years or so; demarcation between conversion during the early stages and later in life is difficult as, even if proper precautions are taken, some rise in temperature during early hardening is inevitable. Moreover, one could argue that if the claim of poor workmanship was not raised at the time of construction it should not be adduced twenty years later, and we feel that deterioration manifest at a late stage is an incontrovertible proof that, under the given conditions, the long-term performance of HAC concrete is inadequate.

We would go even further. If poor workmanship, in the broadest sense of the word, did indeed exist but was not realised at the time, then the standards expected are unrealistic. In other words, we require of the concretor a level of performance which is difficult to achieve on site or even in a factory.

Bearing this situation in mind, we may find it useful to give brief details of various failures of structures made with HAC concrete.

5.2 Structures in France

Since HAC was developed and first used in France, it is reasonable to start with the examples of failures there. Table 5.1 gives a summary. Trouble of the type described under item 1 of the table, although clearly undesirable, is not dangerous as the existence of bad concrete becomes apparent at an early age and remedial action can be taken even before the structure has been commissioned[4]. Moreover, we now know that sea water should not be used as mixing water with HAC (because chloro-aluminates are formed) so that this source of trouble can be readily avoided. We know also that hot water, even fresh, should not be used in mixing, and a combination of salinity and high temperature obviously multiplies the ill-effects.

It is only fair to add that, in contrast to the data of Table 5.1, a number of structures built in France with HAC remained intact, notably port works at Calais built in 1925, at Le Havre in 1927, and at Lorient in 1929.

5.3 Structures in sea water and piles

Hot climates

A number of failures of HAC concrete structures in contact with sea water has been observed in the then French territories in Africa[51]: a pier at Sfax, Tunisia, built between 1924 and 1926, was found to have badly deteriorated by 1933; marine works at Djidjelli, Algeria, constructed in 1938 started to fail in1941; and wharves at Point Noire, Congo Brazzaville and at Libreville and Porte Gentil, Gabon built in the years 1927 to 1930 showed serious damage by 1932, and at the age of ten years were practically completely destroyed. It is possible that unsatisfactory grading of the aggregate contributed to failure but the common factor in all them is HAC.

A discussion[52] published in 1946 reports the failure of the Shab Baraya light in the Red Sea, constructed in 1936. The concrete was mixed with fresh water and was placed in sea water just below the water level. The operation took place during the hot season so that the concrete ingredients were hot and the concrete itself was cured in warm sea water, probably not below 29°C. Some erosion at the water-line was observed 7 months after placing; at 17 months this extended, in places, to a depth of 150 mm. Four years after placing, a cavity 500 mm deep was reported, and the concrete was generally so soft that it could be picked out by hand. The light had then to be dismantled. Some works at Port Sudan also failed, in some cases partially, in others completely, the concrete becoming soft and weak.

Table 5.1 Deterioration of HAC structures in France

Item	*Location and date of construction*	*Condition of the concrete when inspected*	*Remarks*
1	Plougastel bridge on the Elorn, near Brest, 1925 - 28	In a section of the bridge constructed in hot weather, with sea water used as mixing water, abnormal hardening was noticed the day after placing. 56-day strength: 7.6 N/mm^2 1 year strength: 12.4 N/mm^2 Concrete was soft and had a characteristic brownish-yellow colour; on drying, colour changed to dirty light grey.	Engineers: Freyssinet and Coyne[48]
2	Structure near Brest, reconstructed with HAC concrete, 1924	Gradual failure took place until rebuilding was necessary in 1930. (Portland cement was then used.)	
3	Bridge over the Loire, 1933	Concrete satisfactory when placed; no excessive development of heat took place. 1947: strength of concrete in the piles and abutments (i.e. parts in contact with or in the vicinity of water) was 2.1 N/mm^2. Concrete was very soft and dark brown[49].	Conversion did not occur during the initial hardening: had this been so, it would have also taken place in the higher parts of the structure of similar dimensions. The damage was considered to be due to *"a change in the structure of the hardened cement"*, and the cement manufacturers were held to blame and had to pay damages.
4	de la Corde bridge over the Penzé near Carantec (Finistère), 1927	1937: concrete in abutments and in the part of the arch in contact with the sea was relatively hard on the surface but soft and weak in the interior. The stability of the bridge was threatened and reconstruction was necessary.	Chemical analysis of the concrete showed appreciable amounts of magnesia, SO_3, chlorine, and alkalis, which probably came from the sea water, although the manner in which this took place is unknown. In any case, according to Cavenel,[50] chemical attack was not the cause of the deterioration. In our opinion, it is possible that the ingress of sea water followed the increase in permeability caused by the conversion of the calcium aluminate hydrate, the surface appeared comparatively hard owing to carbonation.

5	Marine structure at the mouth of the Gironde, 1928	There is no doubt that the standard of workmanship in construction was high[4]. Concrete remained in a good condition for at least 4 years, as shown by test cores. 1934: concrete between the tide marks was soft; aggregate could be removed by hand[4]. Above the water mark, concrete lost some strength and changed colour to yellow.	Deterioration took place with time and was not due to faults during placing.
6	Pointe de Grave in the same area as 5, 1927 — Landing Stage	1931: sound; 1934: deteriorated	
	Structure in the open and not in contact with sea water	1934: failed	Deterioration occurred only some time after placing.
7	Pier at St. Malo, 1932	1935: intact; 1939: strength of concrete decreased	Deterioration occurred only some time after placing.
8	Landing stage at Grand-Quevilly in the estuary of the Seine, 1926	1931: deteriorated, damage being ascribed to faulty workmanship. Repairs executed under strict supervision but damage recurred several years later and extended to parts of the structure remote from the river.	Cement manufacturers were held to blame. This was so also in the case of a bridge over the Gironde at Pauillac.
9	Landing stage at Port-en-Bessin, Calvados, 1923	No deterioration until 1929 when damage was first observed.	Deterioration occurred only some time after placing
10	Two railway bridges on the line Toulouse-Bayonne, 1927	1941: deterioration first observed.	Deterioration took place a considerable time after placing.

In connection with this case, we may note that the cement makers expressed an opinion that *"some HAC behaved erratically"* when the temperature was in excess of 29°C *continuously* but not so when it dropped during the night. We now know, of course, that temperature effects are cumulative (see Section 3.2) but the views of the manufacturers are of interest as the codes of practice so often refer us to them (see Page 78).

Cooler climates

Let us now turn to experiences in sea water in a cooler climate. Long-term exposure tests on reinforced concrete piles were conducted by the Sea Action Committee of the Institution of Civil Engineers[38], starting in 1929. The piles, 127 mm square and 1.52 m long, were placed at Sheerness dockyard and also at the Building Research Station in synthetic sea water of triple strength.

It was found that HAC concrete piles made with a 1:5 mix and a 51 mm cover to the reinforcement remained uncracked after 23 years' exposure, although some corrosion of the steel was found in about one-half of the piles when broken open. With the same mix and a 25 mm cover to the reinforcement, only some of the piles remained uncracked after ten years but even those cracked later. In the case of a 1:2.6 mix with a 25 mm cover, no cracking was apparent after ten years, and most of the piles survived the full span of 23 years; in these, some corrosion of the steel could be detected. This performance was not equalled by piles made with any other cement.

Piles made with a lean mix (1:9) were all cracked after seven to ten years when the cover was 25 mm. With a 51 mm cover, even these lean-mix piles survived the exposure at the Building Research Station but those at Sheerness nearly all cracked, probably owing to the effects of tide. (At the Building Research Station an artificial tide effect was produced two to three times a week.)

The durability of piles involves a number of factors and the problem as a whole is outside the scope of this book, but some of the conclusions of the investigation are of interest.

Although disintegration of concrete as a direct result of chemical attack by the sea water did not take place to any appreciable extent, in general a reasonable agreement was found between the resistance of the cement of a given type to attack by sea water and the degree of protection against corrosion afforded to the reinforcement. Thus piles made with HAC proved superior to those made with other cements.

The strength of HAC concrete determined by tests conducted concurrently with the investigation on piles[38] is of interest. Table 5.2 compares the strength of 76mm diameter by 152mm cylinders stored in fresh water with that of similar specimens in triple sea water. For mixes with water-cement ratios between 0.29 and 0.53, the latter storage does not appear to have an adverse effect. The half-immersed specimens seem weaker than those fully immersed; our statistical tests indicate that the difference between the two conditions of storage is significant. Concretes with water-cement ratios of 0.85 and 0.93 disintegrated completely in sea-water but even in fresh water their strength was low. It may be recalled that, for all mixes, storage in fresh water resulted in a loss of strength between the ages of five and ten years, as shown in Fig. 4.1.

The piles used by the Institution of Civil Engineers in their investigation were only 127 mm square in section, and the report[38] ends with a warning that *"the risk of high temperatures being attained increases with the cross-*

Table 5.2 **Ten-year strength of HAC concrete stored in water[38]**

Cement	*Cement-aggregate ratio*	*Water-cement ratio*	*Compressive strength (N/mm²) for storage in:*		
			fresh water	*triple sea water*	
				fully-immersed	*half-immersed*
X	1:2.6	0.31	46.6	56.5	48.3
	1:5	0.46	39.5	43.1	42.8
	1:5	0.53	36.1	41.6	38.8
	1:9	0.93	13.3	*	*
Y	1:2.6	0.33	50.0	49.3	47.0
	1:5	0.45	46.2	52.5	47.0
	1:5	0.48	53.2	58.6	56.4
	1:9	0.85	15.2	*	*
Z	1:2.6	0.29	57.2	51.7	51.6
	1:5	0.45	51.9	56.0	51.7
	1:5	0.48	56.2	56.6	56.2
	1:9	0.85	15.1	*	*

* Deterioration so far advanced that no test was possible

sectional area of the pile" and *"the behaviour of this (high-alumina) cement can be much affected if a high internal temperature is reached in the concrete."* Finally, there is a categorical statement that *"high-alumina cement is unsuitable for use in tropical waters."* This is supported by some tests at Sekondi Harbour in Ghana, where a disintegration of piles made with the 1:5 mix and a 25 mm cover was observed within four hours of placing[38].

An investigation[53] of HAC concrete piles standing in sea water showed that after 30 years the strength of the concrete below the water level was unimpaired but the concrete exposed to the atmosphere in South-East England showed a loss of strength. This loss was 10 to 20 per cent at a water-cement ratio of 0.3, and 40 to 50 per cent at a water-cement ratio of 0.48.

Piling

Piling is very important in construction and the possibility of using HAC in piles is of interest. Some further experiences should, therefore, be reported. Piles of a well-known building in Stockholm built in 1928 with HAC concrete have seriously deteriorated after a number of years, and during reconstruction fifteen years ago the strength of concrete was found to be about 4 N/mm^2, compared with 32 to 42 N/mm^2 at the age of four days. Likewise, parts of a bridge, destroyed through a road accident had a strength of 4.5 to 5.9 N/mm^2, although the eight-day strength (in 1933) was 35 to 39 N/mm^2 [54].

In 1949 there appeared a report[55] on experiences with piles 381 mm square and 18.3m long required for a plant in Iran. To avoid casting at a high temperature, the piles were made in Great Britain and were shipped abroad. On the recommendation of the cement manufacturers, a 1:2¼:3 mix with a water-cement ratio of 0.50 was used. This is a water-cement ratio which would not be considered unacceptable even today. The piles were shipped but when the time came to drive them (six to eight weeks after casting) they were found to have deteriorated so badly that they had to be discarded. The strengths of the concrete used were as follows[56]:

preliminary cubes at 7 days	63.5 N/mm^2
works cubes at 7 days	44.1 N/mm^2
cubes sawn from piles after arrival in Iran	14.3 N/mm^2
cubes sawn from piles left in England	29.0 N/mm^2

Many of the piles in Iran were so soft and weak that their strength was believed to be less than 14.3 N/mm^2 as, according to the report[55], *"at such a strength it should have been quite possible to drive the piles"*, and yet, *"it was impossible to drive any of them successfully"*.

The decrease in the strength of the piles left in England is of especial interest as it indicates that, despite precautions, piles cast in England can be subject to an appreciable loss of strength. It is likely, however, that only the piles cast during summer were affected, as some of the piles made in September which had been left in England were successfully driven in Dagenham dock 18 months after casting[57].

We are told that since the Iranian experience the very large company involved has not used HAC in structural concrete except for temporary works.

Some of the difficulties may be avoided by driving the piles at a very early age, before appreciable conversion has taken place. For instance, in the same discussion[58] information was given on a 406 mm square pile (using a HAC 1:1¾:3½ mix) when the air temperature was 27°C. The side shutters were struck 1½ hours after casting when the temperature in the concrete reached 41°C. The pile was driven successfully 26 hours later. Nevertheless, judging from the experience in Iran[55], conversion would still take place, and in case of strongly aggressive ground waters the concrete should achieve a reasonable maturity before the piles are exposed to attack[30].

An English experience with HAC concrete piles was reported in the discussion[98] of our 1963 paper. An attempt was made *"to drive precast concrete piles through hard ground for a bridge foundation near Maidenhead. Here, in spite of the presence of the cement makers' representatives, and the following out in detail of their instructions as regards curing, shutter striking and so on, it was not found possible to drive the high-alumina concrete piles. In every case they broke up under the pile hammer in a brittle fashion. The cube strength at 24 hours was of the order of 8000lb* / sq. in.(55 N/mm^2). *The attempt to use these piles was abandoned. No difficulty was experienced in driving pre-cast piles made with rapid-hardening Portland cement through the same ground. The aggregate used for both types of pile was very similar."*

There have been also cases of successful use of HAC concrete piles. One of these is the Boscombe pier in Bournemouth. This pier is supported by 72 piles of octagonal cross-section, 380 mm across the flats. The piles are 15.3m long, of which 3.1m is below sea-bed level. The piles were precast on site in 1925-1926, using approximately a 1:5.5 mix with an unknown water-cement ratio. In 1935, the pier was inspected visually and found to be in very good condition. During the Second World War, as an anti-invasion measure, five of the piles were partially demolished. They were extracted in 1958-1960 and were found to be in excellent condition. The remainder of the piles are still in service. Likewise, the HAC concrete piles of the Fingringhoe bridge in Essex, driven in 1923 (a good vintage year) are still in service; however, the strength of concrete in the structure steadily decreased with time, as shown in Fig 5.1. There must be many similar cases.

Figure 5.1 **Relationship between the strength of concrete in the Fingringhoe Bridge and age.**

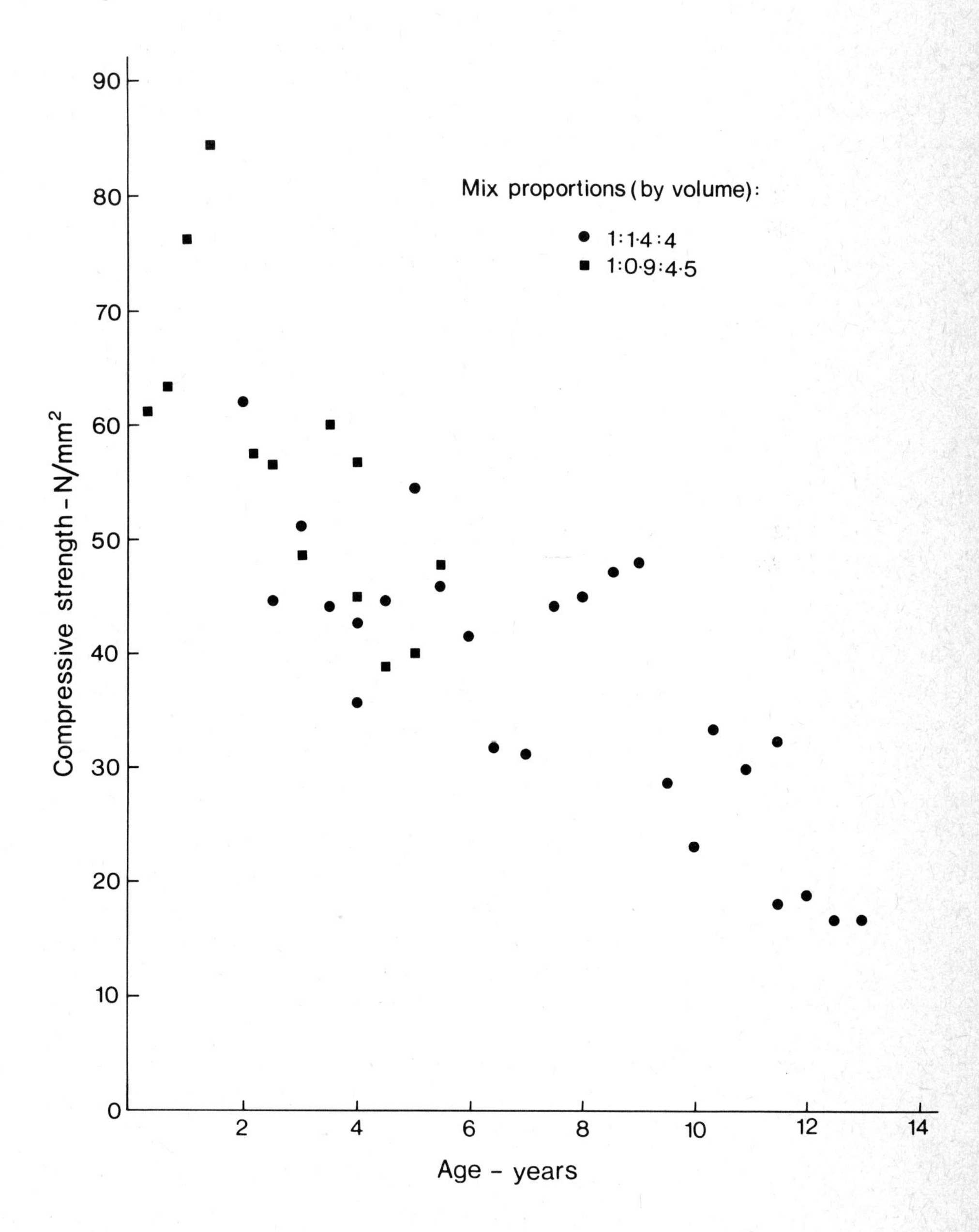

Table 5.3 Deterioration of HAC concrete structures of various types

Item	*Location and date of construction*	*Condition of the concrete when inspected*	*Remarks*
1	Precast prestressed I-beams, 1067 mm high with 203 mm web, and flanges 229 mm, and 127mm thick, cast in sections about 3m long, for a cantilever canopy over a football stadium, in Lancashire, June, 1960 Some of the beams made with HAC; water-cement ratio of 0.58	During post-tensioning, several days after casting, when a stress of 14.4 N/mm^2 was applied, one of the sections failed in compression.	No direct measurement of strength was made but comparative measurements of ultrasonic pulse velocity between the beam that had failed and other beams, some of them already prestressed, led the Author to advise against the acceptance of the beams made with HAC.
2	Base for a crane track in a ship repair yard in London docks, consisting of a 229 mm slab of a 1:2:4 HAC concrete overlaid by 152 mm of a 1:2:4 ordinary Portland cement concrete	Nine years after placing, the HAC concrete was found to have disintegrated and caused settlement, the Portland cement concrete remaining sound. Differential thermal analysis and X-ray diffraction indicated almost complete conversion.	This case is of special interest as the slab of HAC concrete was extremely thin and the position of the slab was such that no rise in temperature above normal could reasonably have been expected at any stage. The soil on which the slab rested contained a small amount of sulphates and an appreciable amount of chlorides, and both of these were found in the concrete; this is rather similar to the behaviour of the foundations described in item 4 of Table 5.1
3	A reinforced concrete chimney built in England using a 1:2:4 mix with HAC in the top 6m and ordinary Portland cement below. The chimney served a coal-fired boiler and was brick-lined. The temperature of flue gases was about 132°C and the temperature at the surface of the concrete was approximately 43°C, and condensation took place in the chimney.	Nine years after construction, the HAC concrete was found to have a strength of 9.6 N/mm^2; a large degree of conversion had taken place, and chemical analysis indicated some attack by the sulphates. The concrete made with Portland cement remained sound.	A similar stack, discharging acid fumes at room temperature, has not been adversely affected in either the HAC or ordinary Portland cement concrete.
4	Vertical colliery shaft in the Midlands, 7.3m in diameter, lined with HAC concrete 457 mm thick. The lining, 274m long, was constructed in lengths of about 24 m, each length being cast continuously through watered ground of high sulphate content, 1960	The outer face of the concrete was placed against frozen ground and the inner face was subject to a shaft temperature of about 50°C. The inner face was sprayed continuously with water but, nevertheless, due to the heat of hydration, a rise in the temperature of the concrete took place. Many months later much of the concrete was found to be weak and friable and the shaft had to be relined with cast-iron.	

5	Foundations of an industrial plant in sulphate-bearing ground in the Midlands; 1:2:4 mix; 1946	Nine years after construction, the concrete in some of the stanchion bases was soft to a depth of about 50mm but sound in others. The samples of the damaged concrete all showed a varying degree of conversion. In most cases chemical attack by sulphates had also occurred but it was not possible to say whether this was the result of conversion or was due to the apparent excessive leanness of the mix. Such a mix would be porous and liable to physical attack by crystallization of salts in the pores but a determination of the leanness of a mix after damage has occurred is difficult; it is not possible to say a posteriori whether the mix was too lean when placed or whether it appears lean due to leaching out of cement from the attacked concrete.	As a remedial measure a drain was put round the building to keep the foundations dry and no further deterioration has been observed. In further work sulphate-resisting Portland cement was used. N.B. : a lump of Portland cement (which was probably a residue from a mixer tipped at the end of a day's work) within a few feet of the faulty foundations was in perfect condition, but this does not imply the superiority of ordinary Portland cement under the given conditions of exposure.
6	Concrete haunching and benching in a large gas works sewerage system in England. Temperature of the effluent is about 38°C.	Disintegration over a period of 4 years. Differential thermal analysis and X-ray diffraction indicated conversion of the hydrated calcium aluminates, and chemical analysis showed that chemical attack had taken place.	
7	Jetty on the south-east coast of England	33 years after construction, a sample of concrete taken from the part between the water marks was found to be very friable but differential thermal analysis failed to prove conversion; it is possible, however, that this preceded the chemical attack, which was evidenced by analysis.	
8	Lining of a water supply tunnel for San Francisco, 1927 (thickness of concrete 762 mm)	A rise in the temperature of the concrete to 82°C was observed and the strength was only a few N/mm^2 [70]	
9	Precast beams used in construction of the soffit of penstock intake structure in a dam in Otago, New Zealand, 1956	The elements were cast on a hot humid day, the air temperature reaching 38°C. The temperature inside the concrete reached 60°C, and the strength of the test beams was found to be 50 per cent of the strength of similar beams kept at 21°C.	This strength would have been adequate from the structural standpoint but, because of the danger of corrosion of reinforcement, the beams were rejected.
10	Railway overbridge, Austria	Early strength of concrete: 59 N/mm^2. Owing to the action of wet steam, the concrete was reduced to a slurry.	

5.4 Other HAC concrete structures

In this section, it is proposed to describe and discuss some of the older failures of HAC concrete structures and in the succeeding chapters to deal with the British failures in 1973 and 1974.

Tunnel in Iraq

Construction of a tunnel through gypsiferous ground in Mosul, Iraq, where lining with HAC concrete was used, led to numerous difficulties[29]. What is of interest in the context of this book is that, although the Building Research Station tests on the influence of temperature on the strength of HAC concrete were published in 1933[28], as late as 1939-40, *"the manufacturers of HAC had given an assurance that the high temperatures prevailing in Iraq would have no ill effects on the concrete*[29] ".

We may note that, despite this experience of the manufacturers' advice, which we discussed in 1963[45], the Code of Practice on the Structural Use of Concrete CP 110, published in November 1972 says:*"At temperatures above 30°C, conversion may occur with a considerable degree of strength reduction, depending on the water-cement ratio, even in concrete which was originally properly cured. The use of high alumina cement in such conditions is not recommended without prior reference to the manufacturer."*

Let us revert to the construction in Iraq. After its completion, the conclusion was drawn that *"when temperatures exceed 85°F (29°C) HAC work involves undue and meticulous control of all operations"*[29]. We can interpret the situation in a different manner: it seems that the strength of the concrete made on the site (approximately a 1:3:6 mix with a water-cement ratio of 0.75 to 0.90) was a direct function of the ambient temperature and increased as the weather became cooler; this is shown in Fig. 5.2.

Figure 5.2

Strength of concrete cubes at the age of 1 to 2 months as a function of mean maximum monthly air temperature.

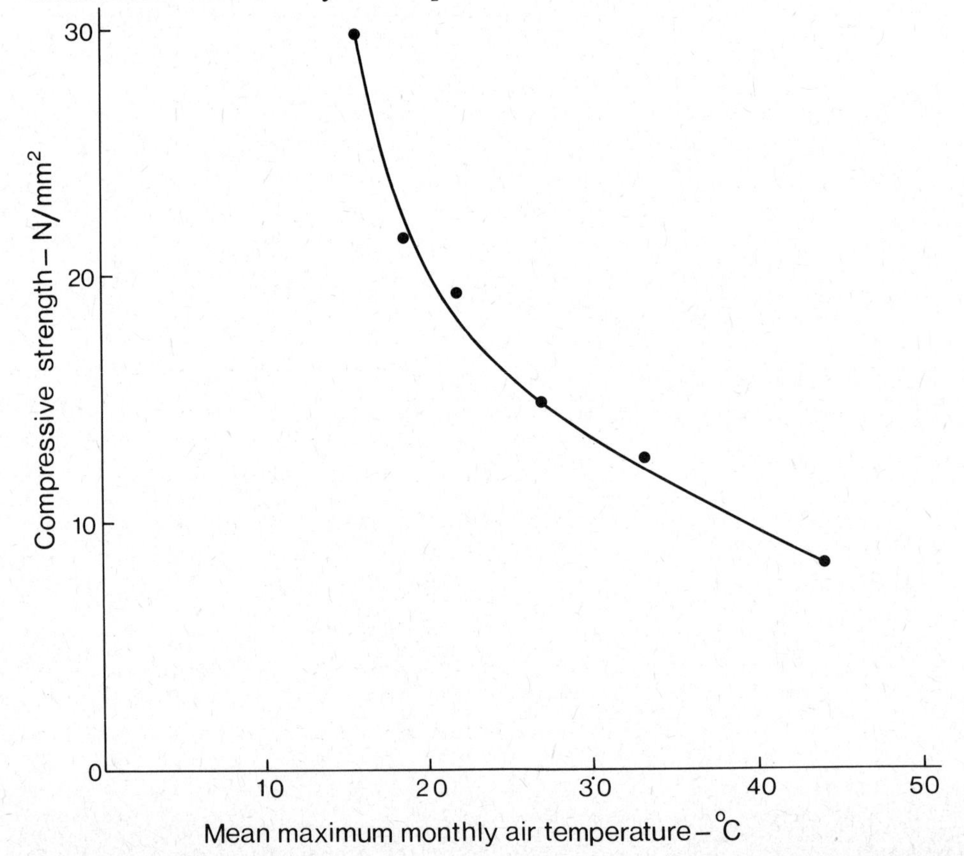

Instances of deterioration of various types of structures are collected in Table 5.3.

Cooling towers

A further case of adverse experience with HAC concrete in England concerns the construction, in 1954, of cooling towers using precast prestressed units. The contractor's scheme was economical only if formwork could be re-used every 24 hours and, since a strength adequate for prestressing had to be developed during that time, HAC was chosen. It was only after many thousands of units had been made that the contractor learnt that under the operating conditions (continuous jets of water at about 38°C) deterioration of the concrete would take place. He then tested 152 mm cubes of concrete similar to that used in the structure by storing them continuously in water at 49°C for up to three months, and found a drop in strength from 55.2 to 5.5 N/mm^2. As a result of these tests, HAC concrete units were replaced with similar units made with rapid hardening Portland cement and steam cured; these proved satisfactory after five years' use[59]. It is clear that the financial loss involved in the original use of HAC was substantial.

It is interesting to note the fate of the HAC concrete units in the exterior and the bottom of the tower, where no high temperature was expected, and which were not replaced but were painted with chlorinated rubber. At the end of four years some of these were soft and weak and had to be replaced but others remained in good condition.

A contributor[98] to the discussion on our 1963 paper[45] wrote that he *"had experience of high-alumina concrete both as engineer and contractor and in every case this experience had been unfortunate, though perhaps also fortunate in that deterioration or failure took place before the associated structures were loaded. Before the war he had used it on concrete piers and small anchor blocks for hydro-electric pipelines in Scotland and North Wales. In both cases, change of colour and progressive deterioration took place very rapidly."* He also added that: *"It was disturbing to note that manufacturers still pressed the sale of high-alumina cement without drawing proper attention to the risks attendant on its use."*

Tunnel in East Anglia

An unusual problem arose in 1969 during the construction of a very large tunnel to carry water in East Anglia. The concrete segments for the tunnel lining were cast in moulds made of reinforced HAC concrete. The mix proportions of the HAC concrete used in making the moulds were 1:1.57:3.33, with a water-cement ratio of 0.40. The aggregate was crushed HAC clinker, known commercially as *Alag*. A strength of 76 N/mm^2 was easily achieved.

The moulds were arranged in a battery and had curved faces. Because of their rather complex shape, concrete moulds were cheaper and faster to make than if they were of steel or timber. It was planned to re-use the moulds a number of times. Curing was by heating in an enclosed space.

The moulds in use were found to be soft in places, including the curved face, to such an extent that the matrix could be easily scratched out and the bond between the matrix and the aggregate was very weak. The cement makers' advice was sought. They expressed the view that a possible cause of damage was the handling and vibration of the moulds, saying that *"the forces engendered by vibration or possibly mechanical shock of some sort may create sufficient tensile stress to fracture even the strongest concrete"*. They considered also the possible effect of the release agent on setting and hydration but it did not occur to them to mention conversion. Our investigation showed that the loss of strength was due to conversion induced by the intermittent application of a high temperature during the casting of the tunnel segments. A calculation of the heat generated by the hydration of the Portland cement in the tunnel segments (cast within the HAC concrete moulds) showed that conversion would be induced, even if none took place when the moulds themselves were cast. A rise in temperature of the moulds during the curing of the segments was estimated to be about 40°C. As a result, parts of the mould in contact with the segments would reach a temperature of 50°C to 65 °C. This situation was aggravated by the external heating.

The moulds had to be discarded and they were replaced by moulds made with ultra rapid hardening Portland cement. A great deal of money was wasted, and clearly HAC should not have been used in the first instance.

All this should, of course, not have happened and the contractor should have been advised against the use of HAC but the theme of ignorance of the properties of HAC is recurrent. A dairy floor in New Zealand was relaid using HAC concrete but was soon found to deteriorate. Attack by lactic acid was suggested as the cause but, as the area in question was used for washing out of milk churns and there was a continuous flow of hot water over the floor, we had strong reasons to suspect that it was the high temperature of the water flowing over the concrete that was the real cause of failure.

Erroneous explanations have been offered also in other cases. For instance, a bridge over the Loire built in 1932-33 was demolished by the French Army in June 1940 and the concrete was subsequently (in 1943) found to be weak and to contain cubic calcium aluminate. It was suggested that conversion was triggered off by the explosion but tests on samples subjected to explosive or ultrasonic waves have not confirmed this supposition [4].

The Bavarian failures

In 1961, the roofs of six agricultural buildings in Bavaria, made of precast prestressed units with HAC, collapsed suddenly, fortunately without loss of life. The units were manufactured between 1953 and 1957. They were thus 4 to 8 years old, which agrees, for instance, with the indications in Fig. 4.5 on the age at which a loss of strength takes place. There were ten other unpublicised floor failures[69] and some other cases of serious deterioration of prestressing wires in HAC roofs in Bavaria; all of these were limited to byres, schools, cinemas, and meeting halls; in total, there were 40 cases[63]. The failures occurred without warning: there was an amusing case of a farmer who slammed a byre door and thus brought the roof down.

A thorough investigation[60] ascribed these failures to the use of HAC, the direct cause of collapse being the corrosion of the prestressing steel. It was noted that the factors relevant to corrosion were: the low pH value within the HAC concrete, the increase in porosity of the concrete associated with its loss of strength, and the high sulphide content of the cement. It appears that in some cases cracks in the prestressing steel existed prior to the collapse; indeed, some individual beams could no longer act structurally and remained in position solely through connections to other beams.

The failures which were first investigated had occurred in byres, where the temperature and humidity are high. The failures of these cow sheds were followed by investigations [60] of school buildings and blocks of flats in Bavaria. These showed no serious danger of total collapse but only a probability of failure of individual roof and floor members, all of which were used in composite construction. In consequence, propping and other strengthening measures in all public buildings were ordered in 1963. In 1967, replacement or repair of all HAC concrete in all public buildings was ordered; this was completed by 1971.

As far as private blocks of flats are concerned, where the loads are low, it seems that, although investigations and inspections have been carried out over a period of 13 years, no clear picture of the safety has been obtained. Spot checks continue, in cooperation with building owners, to determine the moment of resistance of, and prestress in, selected beams[63] .

Over 200 roofs from a variety of buildings were removed and individual units were tested at the Technical University of Munich. No regular pattern of deterioration has been established and the only conclusion drawn was that no structure containing HAC concrete can be assumed to be safe[63]. Most of the concrete had a one-day strength of 60N / mm^2, although some had a strength of only 45 N / mm^2; the cement content was usually 400 kg / m^3, but occasionally it was 350 kg / m^3 [63]. The total floor and roof area involved in Bavaria is estimated at 650 000m^2. There is an additional 850 000m^2 in the rest of West Germany. The owners of HAC concrete buildings formed an organization and obtained financial assistance from the government[63].

In October 1962, the structural use of HAC in Bavaria was forbidden. The ban was extended to other Länder[81], and the German Federal specifications do not provide for the use of HAC in reinforced or prestressed concrete. All existing buildings with HAC concrete have to be"made sound".

Our description of the Bavarian failures in 1963[45], provoked a heated discussion. The director and general manager of Pierhead (Mr. J. P. Mitchell)[64] said that there *"was no doubt that the reason for failure was steel corrosion. There was no reported deflection in the ceilings which would have indicated a loss of strength of the concrete".* He then pointed out that: *"The manufacture of these beams varied in two important points from those manufactured by prestressed-concrete specialists in Gt. Britain, when HAC was used.*

(a) The HAC used in Germany was of a chemical composition different from that produced by the Lafarge Aluminous Cement Co. in France and England, and that could well have been an important factor in these particular instances.

(b) The high tensile steel used was hot-rolled, quenched and tempered, unlike the cold-drawn wire which was used by virtually all prestressed-concrete manufacturers in Great Britain. It was recognized that quenched and tempered wire was more liable to stress corrosion than was cold drawn wire."

In the same discussion, the director of the Central Laboratories, Ciments Lafarge, Paris (Dr. G. H. Sadran)[65] said that the German decision not to allow the use of HAC in structures was made *"a priori without any agreed interpretation of the phenomenon*". He found that this *"was almost certainly a case of severe corrosion of the steel tendons"*, and *"in the Bavarian samples (the cubic hydrate) was absent or existed only in traces. On the other hand the normal* $CaO.Al_2O_3.10H_2O$ *was always clearly identified"*. From this he concluded that:*"The Bavarian incidents, regrettable though they were, should really be classed with similar occurrences involving Portland, Portland blast-furnace, or other cements . . ."*

While corrosion was the immediate cause of collapse of the roof beams, the Bavarian government investigation clearly established conversion and considered it responsible for the porosity of the concrete and hence for the increased corrosion.

A strong contribution to the discussion of our paper was made by the former chief engineer of Pierhead Ltd. (Mr. J. Bobrowski)[66]. He said that he *"had been able to discover, by personal investigation on the spot, that the main offenders were German contractors and manufacturers who, with their proverbial precision, applied religiously to HAC concrete the provisions of various DINs (Federal German standards) covering the use of Portland cement, strength alone being the basis for comparison. These attitudes resulted in structural use of mixes with water-cement ratios which were far too high. But even then, where perfect conditions for conversion should have been created, he had found that most of the trouble was due to sulphide corrosion of steel. Hence, considering these failures, one should at least bear in mind that on the Continent a quenched and tempered steel was used for prestressing while cold drawn wire was used in Gt. Britain. In addition, German Rolandshütte brand of HAC contained sulphur impurities, the amount of which was negligible in British and French HAC. Last, but not least important was the fact, especially if considerable lowering of pH value was considered, that German HAC, unlike British HAC, did not have a 6-h strength. The manufacturers therefore resorted to the use of calcium chloride. That last fact would not, of course, be readily disclosed to official state investigators."*

In answer to this statement I wrote in the closure to the discussion[46]: *"Mr. Bobrowski accused the Germans of stupidity ('with their proverbial precision, applied religiously to HAC concrete the provisions of various DINs covering the use of Portland cement'), dishonesty ('resorted to the use of calcium chloride'), and mendaciousness ('the last fact would not, of course, be readily disclosed to official state investigators').*

The German technical press 'seems to incriminate essentially the aluminous character of the cement and its property of conversion' [68] *but both Professor Lafuma*[68] *and Professor Rüsch*[69] *were of the opinion that corrosion of tendons had taken place. However, Professor Rüsch, who lived in Bavaria and was the leading German authority on concrete, had written to the Author in the following terms:*

'1. *As you indicated in your Paper, the* conversion *of HAC increases its porosity. As a result, the aggressive medium can penetrate the concrete more easily.*

2. *Concrete made with HAC does not offer the steel as good a corrosion protection as concrete made with Portland cement, as the alkalinity of the former is quite low. The extensive investigations which we have conducted have shown that the basic reaction is lost after a few years over a thickness of several centimeters, even in concrete which was originally dense and made with a low water-cement ratio. On these grounds alone, the use of HAC in reinforced- and prestressed-concrete structures must be absolutely avoided.*
3. *While the circumstances are not yet quite clear, it can be assumed as definitely established that concrete made with HAC produces favourable conditions for corrosion of reinforcement. The investigations which are being conducted by us aim at a clarification of the circumstances of these processes.'*[69]

Item 1 above showed that, despite Dr. Sadran's vehement assertion, conversion could not be eliminated. Likewise, Professor Lafuma*[68] *believed that it was difficult to establish the part played, among others, by conversion which could have been effectively assisted by the humidity and warmth in the stables.

In connection with the Bavarian failures the Author had stated that: 'It is possible that corrosion of the prestressing wire contributed to failure. Corrosion may be accelerated by the comparatively low alkalinity of HAC but could also be influenced by the increase in permeability in conversion.' This view did not appear to have been shaken."

To close the consideration of the Bavarian failures it is only fair to add that German HAC differs from the cements produced in some other countries, notably the United Kingdom, in its method of manufacture and, what is more important, in its high sulphide content: 1.3 per cent is probably an average figure. As pointed out in a German report[60] published in 1964, the sulphide contributes to the corrosion of prestressing steel, but the denser the concrete the better the protection against outside attack. However, an increase in porosity facilitates the access of CO_2 and lowers the basicity to a critical point at which the corrosion of the prestressing steel, both of the conventional surface variety and by hydrogen embrittlement, is possible. Thus corrosion is due to a combination of the low pH value of the HAC, its increased porosity due to conversion, and a high sulphide content.

Other German investigations[117] showed in 1966 that the pH value of HAC is not intrinsically low but CO_2 leads to the formation of calcium carbonate and alumina, with a consequent lowering of the pH value. The ingress of CO_2 is made possible by the increased gas permeability of the converted HAC[118], so that it is the conversion, and not the sulphide content, that is responsible for corrosion[119]. Indeed, it has been observed that the corrosion of steel in the German HAC concrete occurs only when conversion has taken place[118].

Other failures

Finally, let us report a few more cases of trouble with HAC concrete structures. There were some in Denmark and in Switzerland but the jobs were executed by small contractors and it has not been possible to obtain full data and to evaluate the cause of the deterioration. Among minor applications, the use of HAC in setting fireplace tiles has in many cases led to failure, and tests have shown the conversion of the aluminates to be complete. In Spain, there were several failures of precast HAC concrete beams in agricultural buildings but there were no disastrous effects.

In contrast to the various cases of deterioration, HAC has been used successfully in repair work of a number of structures in the Netherlands, and no trouble has been reported after 17 years. The cement was also used in the lining of the Ericht Tunnel in Scotland in 1928-30 and has stood up extremely well, although in another tunnel lined at about the same time the use of HAC did not prove successful. Many floor and roof beams made in Great Britain with this cement have led to no trouble[36], and there must be numerous other cases of successful use of HAC although we may not know whether or not there has been conversion.

It is reasonable to point out, however, that except in prestressed concrete beams, the actual working stresses in the majority of ordinary structures are so low that failure is unlikely even if full conversion of a high water-cement ratio mix has taken place.

The remarks about the successful structural use of HAC have been made because we do not want to say that it is impossible to produce HAC concrete

Figure 5.3 **Influence of the water-cement ratio on the strength of HAC concrete cured at 100°C[71].**

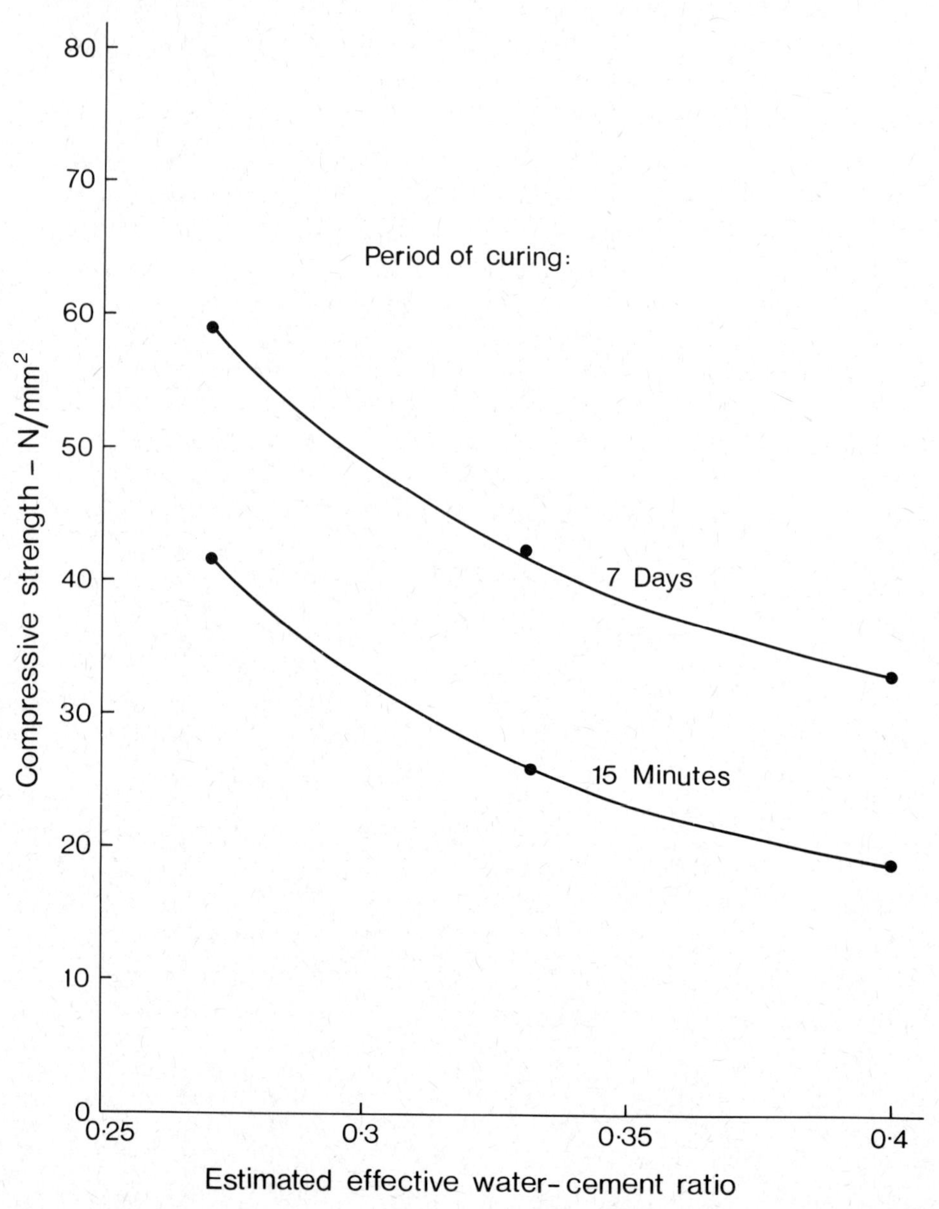

with a long-term high strength. But the conditions which have to be satisfied are onerous and cannot be guaranteed in precast concrete factory production, although they can be achieved in laboratory work. Let us quote as an example some work done by the cement makers, published in August 1971[71]. A very rich mix (1:1.25:2.5) with a maximum size of aggregate of 9.5mm was used. The *total* water-cement ratio was 0.38, i.e. the effective water-cement ratio was probably 0.33. Cubes, 70mm in size, were made, sealed with a greased cover plate, and immediately placed in water at 100 °C. At the end of one hour, the compressive strength was slightly above 30 N/mm^2. With a synthetic cementitious aggregate, made of HAC clinker, the strength reached was even higher: 46 N/mm^2. But the important point is that even a small increase in the water-cement ratio strongly affects the strength; this is shown in Fig. 5.3 from the same source[71].

We may note that the authors of the paper consider the water-cement ratio used by them as high; they say: *"This ratio was deliberately chosen to be rather high (by the standards of HAC concrete subjected to severe conversion conditions), the reasoning being that the concretes would thus show more rapidly any untoward effects of the high-temperature treatment."*

The control in the tests[71] was excellent but they were performed in the laboratory on 70mm and 102mm cubes; could this be repeatedly achieved with the much larger precast prestressed concrete beams made of concrete mixed in a commercial mixer? What about the homogeneity of the mix? What about thermal stresses?

And yet it is the industrial exploitation that the authors recommend[71]. For, in their conclusions, they say: *"In the practical field the results indicate that strong durable HAC concrete can be produced almost instantly by curing at a suitably high temperature, provided that the concrete is mixed using a low water-cement ratio. The industrial advantages this suggests are the potential to re-use moulds far more rapidly than at present on repeat castings, and perhaps the potential to dispense with the mould altogether by using techniques such as extrusion through a heated die to slip-forming followed by heating for a short period." They add that: "In these investigations the mix used was relatively rich to permit hand compaction, but more economic mixes can easily be designated for vibratory compaction."*

What is equally disturbing is the conclusion[71] that *"under mild conversion conditions (i.e. temperate climates), the strength changes accompanying conversion, particularly with low water-cement ratios, may be of little practical significance"*. This chapter perhaps dispels this notion.

We should also deal with the argument that with a strictly controlled water-cement ratio satisfactory HAC concrete can be produced. Experience seems to have shown that this is not so under the usual factory conditions: the 'guaranteed' water-cement ratio is inevitably exceeded from time to time. This could be avoided only under production conditions more appropriate to a pharmaceutical enterprise than to a precast concrete factory.

6. The 1973 Failures

The 'HAC problem' in the United Kingdom built up to its present crescendo through three collapses involving HAC concrete, although this is not to say that, in all cases, it was necessarily the presence of HAC that was alone *the* cause of collapse.

The first of the trio to be described is the roof of the assembly hall at the Camden School for Girls, the second will be the roof of the Bennett Building at the University of Leicester, and the third, the roof of the Sir John Cass's Foundation and Red Coat Church of England Secondary School in Stepney. Let us consider them in turn in some detail, the first two in this chapter and the third one in Chapter 7.

Plate 6.1 **General view of the Camden school assembly hall after collapse.**

6.1 The Camden school collapse

The collapse at Camden occurred shortly after 10 p.m. on 13th June, 1973 (see Plate 6.1). At that time, the building was unoccupied so that there were no casualties. A full report[72] on the occurrence was prepared by the Building Research Establishment. It seems that, about eight hours before the collapse, a mistress who was at the time in the assembly hall heard sounds *"like a shower of gravel falling"* on two or three occasions. Some creaking was also heard. At about the same time, a crack, approximately 0.75m long, was noticed between the ceiling and the edge beam. A girl went to inspect the roof above and noticed *"a sag in the roof surface in the region where the noises had been heard which extended for a distance of 0.6m or a little more"*[72]. The actual collapse was heard by the caretaker but seen by no one.

Structure

The assembly hall was built in 1954-1955 and is shown in Fig. 6.1. The structure was thus about 18 years old at the time of the failure. The relevant part of the hall was almost entirely of precast concrete and the roof consisted of prestressed concrete beams spanning across the hall (see Fig. 6.1). The beams, shown in Fig. 6.2, were of rectangular cross-section, 394 mm by 102 mm, and were prestressed with eighteen 5 mm diameter pretensioned wires and reinforced with two longitudinal bars, 5 mm in diameter,to which there were fixed vertical links, 5 mm in diameter. When the beam was placed in situ these links were bent so as to become embedded in the structural concrete screed. Thus, the links, as well as the castellated upper surface of the beams (see Fig. 6.2), provided a connection between the beams and the screed. At each end of the beam, there was a nib, 220 mm deep and 38 mm long, to provide a seating for the beam. To ensure integrity of the beam-ends a bent bar, 10 mm in diameter, was provided as shown in Fig. 6.2.

Figure 6.1 **General arrangement of the assembly hall at the Camden school[72].**

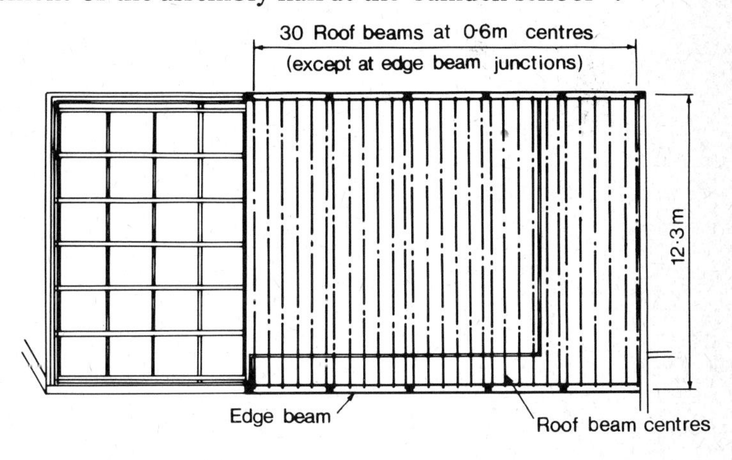

Schematic plan of roof beams

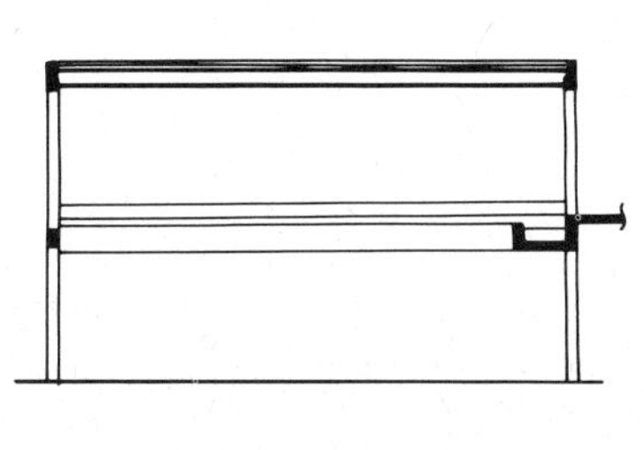

Cross-section

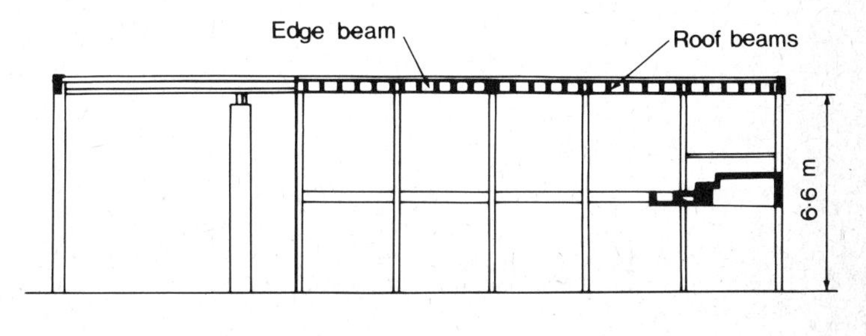

Longitudinal section

Figure 6.2 **Detail of precast prestressed concrete roof beams at the Camden school[72].**

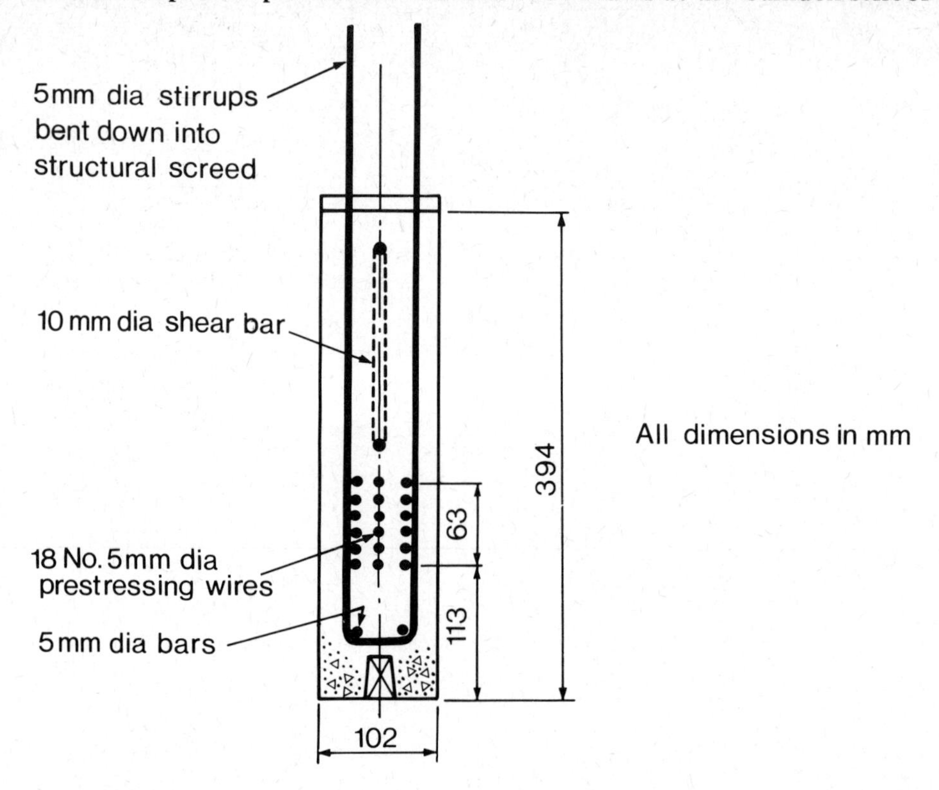

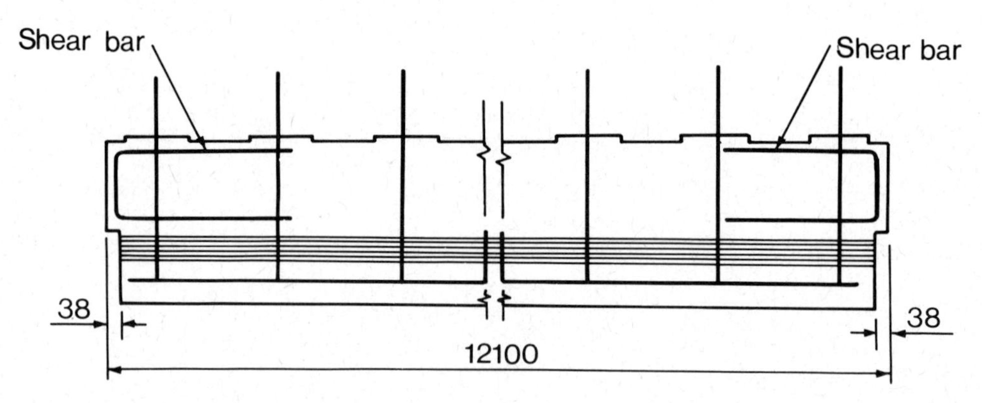

The structural screed had a thickness of 38 mm, increased to 90 mm near the ends of the span, and was cast on a permanent formwork of corrugated asbestos cement sheeting between the beams. The screed was reinforced with a steel wire mesh, 3 mm diameter at 150 mm centres in both directions. The roof was then covered with an insulating lightweight concrete screed, 50 mm to 100 mm thick, and was finished with a 19 mm layer of asphalt. Details are given in Fig. 6.3, which shows also the continuity reinforcement between the edge beams and the prestressed concrete beams through the medium of the structural screed. The roof beams were spaced at 610 mm. Further details of the construction are given in the report of the Building Research Establishment[72].

The collapse

'On the fatal night', the whole of the roof supported by the prestressed concrete beams fell. The longitudinal edge beams, which had supported the roof, remained in position but sustained damage at the majority of the bearings for the roof beams. Initial inspection showed that some of the mild steel continuity reinforcement had corroded badly and that the prestressed concrete beams—the only part of the structure made with HAC—*"showed visible signs of conversion in some places"*[72]. The prestressing steel was in a good condition.

The investigation[72] of the failure led to the conclusion that *"the collapse of the whole roof was probably started by the failure of a single bearing . . ."* This is an important point in that the loss of integrity of a single beam can be seen to lead to a sudden and total collapse.

Figure 6.3 **Detail of roof beam to edge beam connection at the Camden school[72].**

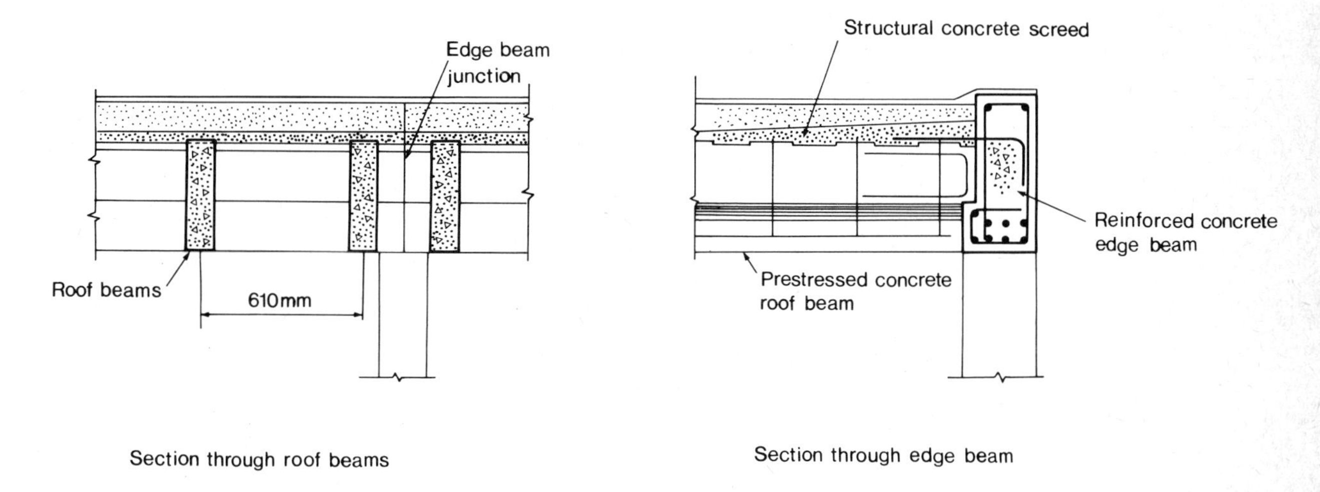

Mode of failure

The mode of failure of this single beam is, as the report[72] says, more speculative but there is some indication that there was a failure of the nib at one end of the beam (see Plate 6.2). It is possible that some slight additional load was transferred to the bearing from the adjacent steel roof owing to the hogging of the prestressed roof on the particularly warm preceding day. *"This increase in load would not seriously overload the beam in bending but could cause fracture at the bearing if its strength were already critical."*[72] It is likely that the noises heard and the movement observed prior to failure were due to the failure of the bearing. Examination showed that the nib had fractured vertically along the line of the shear bar. The damage to the edge beam (made with Portland cement) was negligible. The nominally 38 mm bearing was particularly short (25 mm) and the two continuity bars closest to the beam in question were rusted through before the collapse. Concrete from the fractured nib adjacent to the bearing was found to have converted between 65 and 75 per cent. The report[72] says that this indicated *"some reduction in strength from the maximum attained but probably not to a lower value than that assumed in design."*

Another mode of failure could be postulated: slip at the reduced bearing due to temperature effects and to creep and shrinkage of the prestressed roof beams. However, during the decorating of the assembly hall in the preceding year no movement was reported and there was no water penetration. Loss of prestress by failure of the bond would have caused slip of the wires at the ends of the beams and consequent sagging, but no puddles of water on the roof had been observed. Although laboratory tests on portions of the prestressed concrete beams showed *"that the prestress in some of the beams was less than would normally be expected, the evidence does not indicate that this loss of prestress caused any relative slip at the bearing."*

Plate 6.2

View of the edge beam at Camden school showing on the left the bearing where the nib of the first beam failed and on the right the bearing for the second beam.

Analysis of the failure

Several aspects of the analysis of the failure contained in the report[72] are worthy of note. First, the margin of safety, although not high to begin with, *"was unlikely to have been affected by the conversion of the high alumina cement concrete, since the resistance to bending failure would have been developed between the structural screed of Portland cement concrete and the prestressing wires; it may well have increased with time as the strength of the structural screed increased."* This observation pinpoints the importance of composite construction with Portland cement concrete but of course the composite action must be properly designed and executed.

The second observation concerns the strength of the HAC concrete. Details of the mix are not known but it is possible that a 1:2:4 mix with a water-cement ratio of 0.58 was used. Nothing more definite can be stated. Tests on cores after the failure yielded equivalent cube strengths between 24 N/mm^2 and 42 N/mm^2 with conversion between 77 per cent and 94 per cent. The range of splitting strengths was 2.2 N/mm^2 to 4.0 N/mm^2. Spot checks on the degree of conversion at the fractured nib gave values of 65 per cent and 75 per cent, *"indicating that the strength was probably slightly greater than that assumed in design, viz. 41 N/mm^2. Other values of the conversion of the cement in small specimens of concrete were as high as 95 per cent."*[72]

Two paragraphs in the report merit quoting in full. *"Conversion of high alumina cement in concrete is the result of a change in the structure of the hydrated cement from a metastable form, which develops initially, to a more stable form generally of lower strength. The rate at which conversion takes place is dependent on temperature, the times for 50 per cent conversion being: 15 hours at 50°C, 4.2 years at 25°C and 27 years at 18°C. For concrete with a water-cement ratio of 0.6 maintained in humid conditions for 5 years at 38°C, conversion is approximately 90 per cent and the reduction in strength is about 60 per cent. For a similar concrete maintained in humid conditions for 5 years at 18°C, conversion is about 30 per cent and the reduction in strength is 30 per cent. From these data it is concluded that the degree of conversion experienced in the roof at Camden School could have been attained if a steady temperature of about 25°C had been maintained throughout its life.*

Analyses were made of the temperatures likely to be reached in the roof under various conditions. The results show that on a sunny summer day the temperature in the roof beams could very occasionally reach 25°C and this received some confirmation from measurements in a somewhat similar roof. In the vicinity of the lights with the lights on for several hours, however, temperatures could rise to 25°C fairly frequently and occasionally to 30°C with average winter conditions. These estimates show that conversion of the cement to the extent experienced is consistent with the data referred to earlier. Measurements of ultrasonic pulse velocity through the concrete at points along one of the beams, that had been adjacent to a lighting unit, showed lower velocities near the lights, indicating a greater reduction in strength in that region."

Conclusions of BRE report

Let us now look at the conclusions from the report. The factors which contributed *primarily* to the failure of a joint between the edge beams and the roof beams were:

"(i) insufficient bearing of the prestressed concrete roof beams and the edge beams to allow effective reinforcement of the nibs at the end of the roof beams;
(ii) termination of the prestressing wires within the span of the roof beams without provision for the continuation of longitudinal reinforcement into the nibs above the bearings;
(iv) reduction in strength of the concrete in the prestressed roof beams resulting from the conversion of high alumina cement."

The *subsequent* collapse of the whole roof was mainly due to the following factors:

"(iii) insufficient structural cross-tying of the building;
(v) deterioration and corrosion of the continuity reinforcement between the edge beams and the roof ."

Point (iv) is of particular interest to us. The report[72] says that the *"deterioration of the strength of the concrete has contributed to the failure of the joint although the strength at the critical section had probably not reduced below that assumed for design. Thus, although the concrete was of the required strength, the margin of safety was insufficient. Had the joint detail, however, been satisfactory, the maximum reduction in the strength of the concrete experienced might not have caused failure."*

Summary

There is little we can add to this except to repeat what we said over ten years ago[9] : failure of a concrete structure is rarely due to a single or isolated cause; it is usually a combination of causes that leads to collapse. It behoves us therefore to avoid any potential contributory factor that can be avoided. Without prejudging the case at Camden, it seems reasonable to suppose that, had the bearing been larger, failure may have been prevented. Likewise, had HAC not been used, failure may well not have occurred. It appears that it was the combination of the two, each of which could have been avoided, together with unhappy structural details, that led to the failure.

6.2 The University of Leicester collapse

The building whose roof collapsed was the Bennett Building. The failure took place one day before the Camden school failure, viz. on 12th June, 1973 just before 11 a.m. A student and two cleaners were present in the critical room, the Geography Reading Room. The cleaners left and soon after the student noticed flakes of plaster falling near her and heard some slight noises in the roof; in consequence, she left the room, and immediately thereafter a part of the roof in that room collapsed. The student was lucky as well as quick-witted.

The Author inspected the building some two weeks after the collapse and has conducted some investigations into the failure.

Structure

The Bennett Building was constructed in 1963-1964 and is approximately 82m by 31m in plan, and in the area of the collapse has a basement and two storeys above ground level. The floors are of in situ reinforced concrete flat slab construction. A part of the roof consists of precast prestressed concrete beams which support permanent precast concrete slab shutter panels covered

Plate 6.3 **A detail of one of the pockets which did not fail in the edge beam in the Bennett Building. The crumbling of the seating is evident and has permitted the roof beam to move downwards.**

Figure 6.4 **Schematic representation of the Bennett Building.**

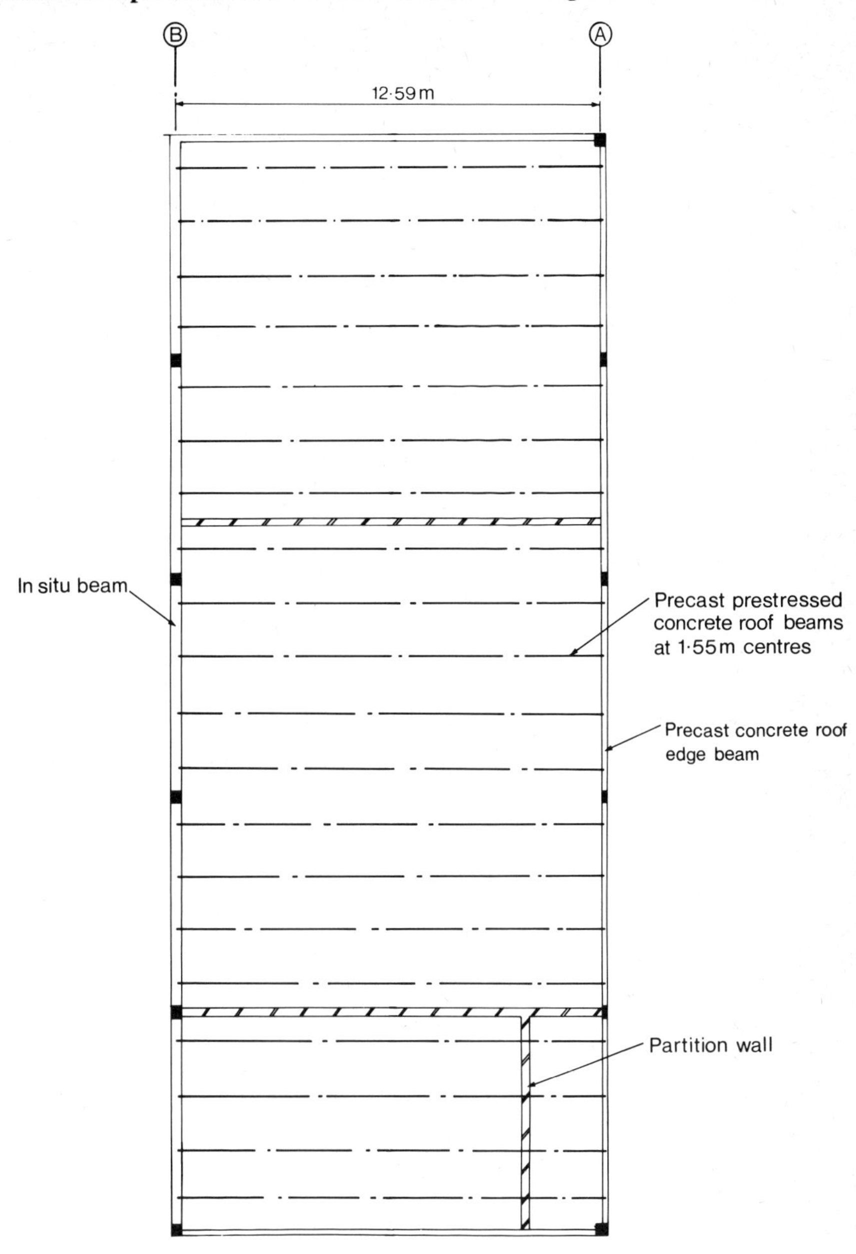

Schematic plan of roof beams

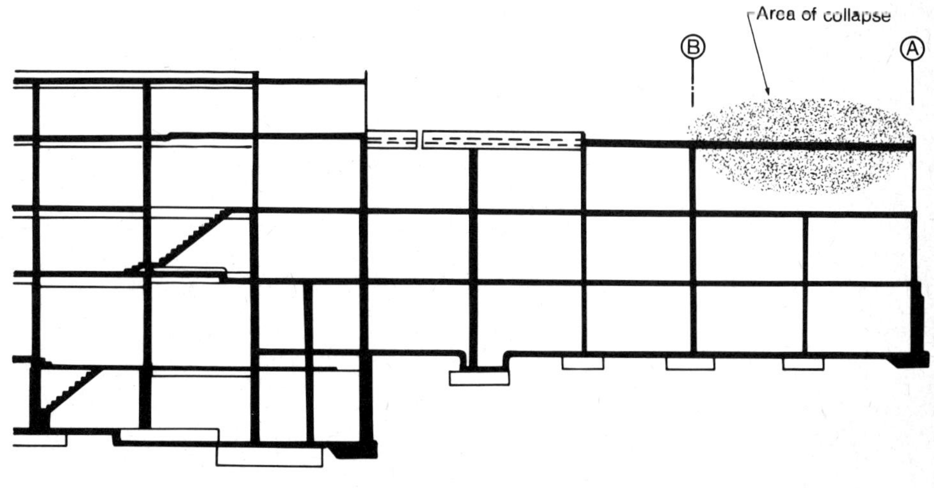

Schematic part-elevation at area of collapse

Figure 6.5 **Detail of roof members in the Bennett Building**

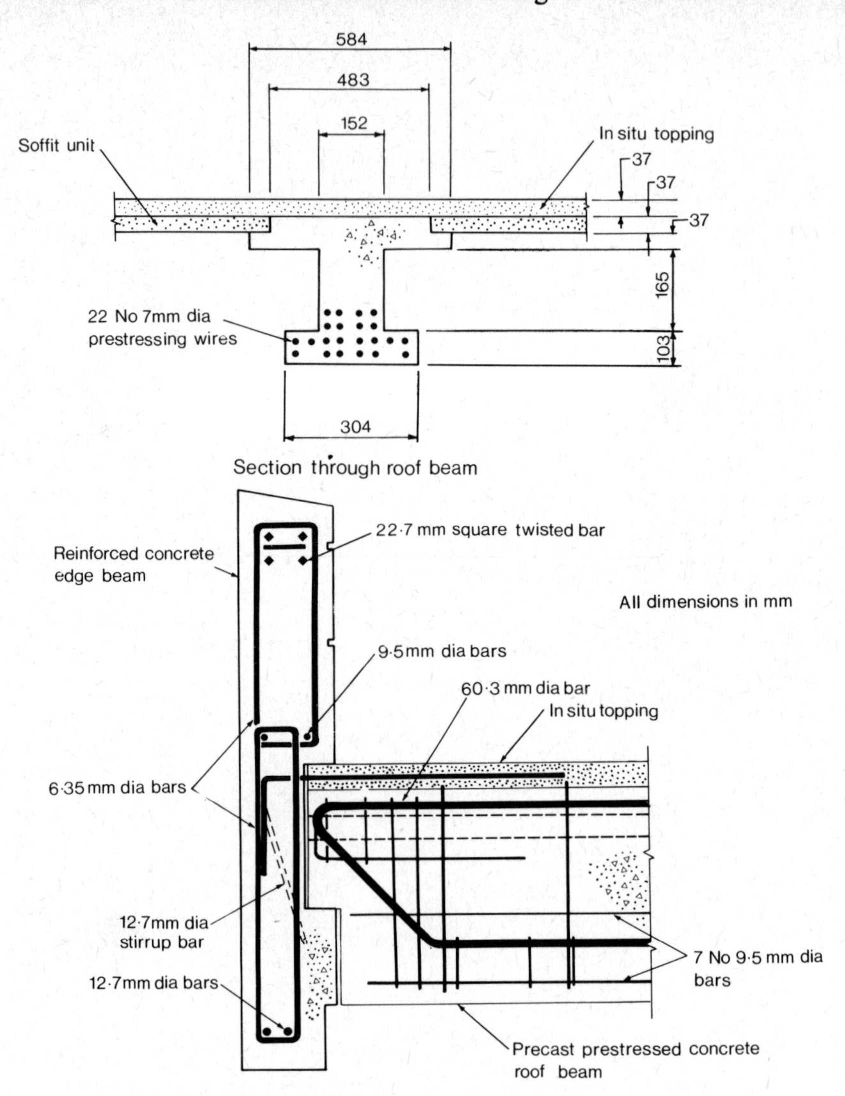

by in situ concrete; the other part consists of in situ flat slabs. The floors and the roof are supported by precast concrete beams and columns. The concrete used was made with HAC.

The area of collapse, shown in Fig. 6.4, is approximately 12m by 6m in plan. The roof beams, shown in Fig. 6.5, were precast prestressed I-section beams, 343 mm deep, placed at 1.6m centres, and spanning 12.5m. The beams were supported at the grid line A of Fig. 6.4 by 50 mm deep pockets formed in the precast concrete edge beam and at the grid line B by an in situ concrete beam.The in situ beam was cast against the ends of the prestressed concrete beams. Fig. 6.5 shows the details of the precast soffit units, 38 mm thick and spanning 1.0m between the flanges of the prestressed beams. In situ topping, 38 mm thick, was placed on top of the prestressed beams and soffit units to act compositely with the beams; ties to the edge beam were provided. The topping was covered with insulating lightweight concrete roof screed and with asphalt. There were roof lights, about 6m long, between some of the prestressed beams.

The collapse

In the collapse, three adjacent beams fell together with the connecting slab and roof lights; they all landed on the first floor slab. Another beam collapsed partially: it lost its seating but jammed against the roof edge beam and so did not fall completely. It seems that support was first lost by failure of the pockets; the support at the other end remained sound. There was of course consequential serious damage but no collapse of the suspended floor on which the roof landed. Plate 6.3 shows the crumbling of one of the pockets which did not fail.

Inspection after failure showed that the collapsed beams remained connected to the soffit units, to the topping, and to the screed. The beam ends appeared to be very little damaged and there was no corrosion of the reinforcement; the prestressing wires did not show deterioration and the bond appeared to be satisfactory.

It appears that the concrete in the roof beams was highly converted: tests showed a degree of conversion ranging between 80 and 95 per cent excepting one sample which was converted to a lesser degree. However, in all likelihood, the beams were not too weak to carry the loads acting, although the concrete strength was between one-third and two-thirds of that originally specified.

The situation is totally different with regard to the roof edge beam where severe deterioration of the concrete occurred. We understand that in some cases cores could not be taken intact but crumbled during the coring process. All the cores were found to be highly converted.

Since the section below the supports to the prestressed beams was unreinforced, it is possible to visualize a situation where the supports could fail when a sufficient loss of strength had occurred. It is reasonable to speculate that the use of HAC in the edge beam may have reduced the time interval between the completion of the structure and failure. Had Portland cement been used, then the deterioration of the seats may have occurred more

Plate 6.4 **Steel joists and columns providing permanent support in the Bennett Building.**

Plate 6.5 **As Plate 6.4 but the column partially encased in asbestos.**

slowly. Indeed, such deterioration was observed in another building constructed with Portland cement.

Strength data

Some data on the strength of the roof edge beam are of interest. The cube strength dropped from a specified value of 55.2 N/mm^2 at 7 days to an equivalent cube strength (estimated from cores*) of between 6.9 N/mm^2 and 20.5 N/mm^2. the degree of conversion ranged from 85 per cent to 95 per cent. It may be noted that the least strength measured was 13 per cent of that specified.

These very low strengths make one suspect a poor quality concrete mix or bad workmanship. There are indications that the ratio of fine to coarse aggregate used was 2:1 instead of the usual 1:2. The latter appears in the original specification and in the contractor's mix proposals, and is also confirmed approximately by the analysis of the concrete after the failure. There was also significant segregation in some areas of the edge beam.

An incident during construction is relevant in this connection. In March 1965, the cement makers received a letter from the contractor requesting

* see Section 12.4

them to investigate the cause of the poor concrete in a precast unit which had been cast the previous summer and erected on site. The concrete in the top flange at centre span had cracked and could be pulled away by hand. The cement makers analysed a sample of this concrete and concluded that it had a low cement content and a high water content. The beam in question was removed and rebuilt in situ. It is not known what action, if any, was taken to discover and examine any more beams that may have been cast at that time.

Errors in mix proportions are, of course, difficult to establish with certainty after ten years but we think it likely that similar errors committed with Portland cement concrete would not have been equally disastrous because conversion magnifies the effect on strength of an increase in the water-cement ratio.

Designers' view

Since the cause or causes of failure of the Bennett Building have not yet been established with certainty, it is reasonable to present the view of the designers of the building. Incidentally, it was the same firm of consulting engineers that designed the Camden school building described in Section 6.1 but there was an interval of about eight years between the two designs. The designers, in commenting on another consulting engineer's report, say, in part[75], *"The fact that collapse occurred during a period of relatively high external temperature indicates that the horizontal loads from temperature variation led to failure.*

The stress distribution in the edge beams adjacent to the seatings is difficult to assess. The vertical load produces local high bearing stresses dependent on seating conditions, and, together with the horizontal loads, tensile stresses are set up particularly at the root of the seating pocket.

Two systems of distress could occur. If the bearing stress is high, local crushing of the concrete is possible. There are signs that this may have occurred but to our minds this in itself would not lead to failure of the seatings. Secondly, the tensile stress could reach the ultimate tensile strength of the concrete and cause failure. This has no doubt taken place in at least one seating pocket in the area of collapse. With such a failure, load would be distributed to adjacent roof beams causing further full or partial failure of their seatings.

The main factor, therefore, causing collapse was the inadequate tensile strength of the concrete in the edge beams. Investigations have shown that the strength of the high alumina cement concrete in the edge beams had reduced to a minimum crushing strength of 6.9 N/mm² as against a specified strength of 55.2 N/mm². The loss of strength was due to conversion, and, to some extent, segregation of the concrete below the seatings during manufacture. The preliminary and works cube tests were up to standard, and as the contractor raised no queries on the mix or water-cement ratio, there was no evidence of the concrete not being satisfactory.

From information on high alumina cement concrete published since the structure was designed, it would appear that a greater reduction in strength due to conversion than originally anticipated would have taken place. In fact, the failure at the Bennett Building has highlighted the limited knowledge of the long term behaviour of high alumina cement concrete."

This last comment confirms what we said in Chapters 4 and 5, but we do not agree that the loss of strength was greater than could have been

Plate 6.6 **As plate 6.4 but the column and joists completely encased in asbestos and ready for the finishes to be applied.**

anticipated. In Chapter 4 we reviewed the early reports on loss of strength of HAC; the information has been available for a long time.

And, finally, two of the conclusions[75] are worth quoting as they so clearly support our thesis that there is always more than one cause of failure: "*We agree that if reinforcement had been provided in the concrete under the seatings, collapse would not have occurred but we cannot accept that the lack of reinforcement was a cause of collapse.*

The segregation of the concrete under the seatings during manufacture and the conversion of the high alumina cement concrete reduced its crushing strength to a minimum of 6.9 N/mm^2. The tensile strength of the concrete was so reduced that failure occurred. Had the tensile strength not fallen to well below the estimated value, there is no reason to think that failure would have occurred."

Repair work

That part of the roof of the Bennett Building which has not collapsed has been left in place. The collapsed beams have been replaced by steel joists. The edge beam has been left in place but it has been structurally replaced by steel joists resting on steel columns. The entire roof is being strengthened by the introduction of supporting angle brackets.

In the part of the building which was not affected by the collapse, steel joists and columns have been introduced, as shown in Plate 6.4. They have been encased in asbestos to ensure fire protection. Similar strengthening is proposed for another building of like construction. The total cost may exceed £600,000.

Plates 6.4, 6.5 and 6.6 show typical steel joists and columns and their appearance at various stages of installation up to encasing in asbestos.

7. The Stepney School Collapse

The full name of the school is rather impressive: Sir John Cass's Foundation and Red Coat Church of England Secondary School, and it is located in Stepney, London. The building which collapsed was the swimming pool hall but the adjacent gymnasium plays also a significant rôle in the affair.

The collapse

The collapse occurred shortly before noon on 8th February, 1974, and the absence of casualties was even more miraculous than in the case of the

Plate 7.1 **The Stepney swimming pool after the collapse of the first beam.**

Plate 7.2 **The Stepney swimming pool after the collapse of the second beam and of the roof cladding.**

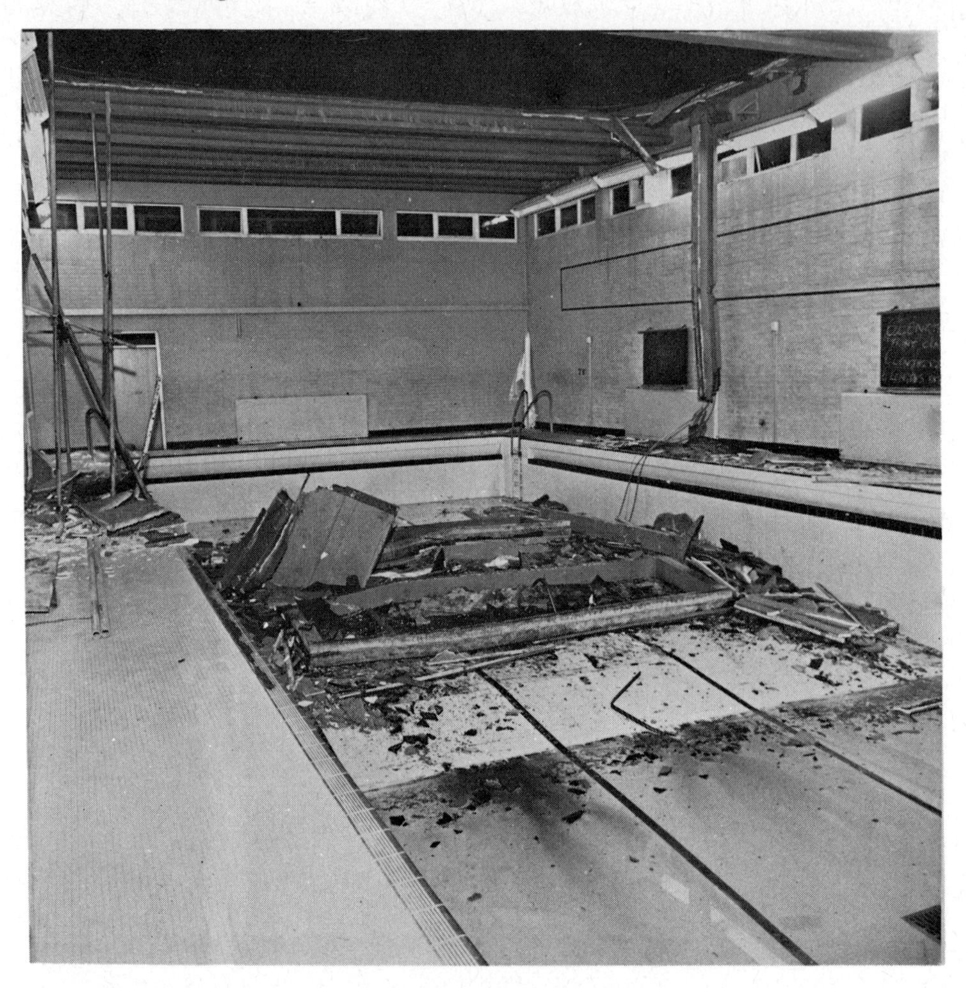

buildings discussed in Chapter 6. Pupils were being taught to swim that morning and one of the girls noticed a fragment of concrete or plaster in the swimming pool. Shortly afterwards, another small fragment was seen to fall into the pool near the same spot. The swimming instructor was alerted and he noticed a crack developing in one of the roof beams and something which he thought to be ballast to be falling from it. He showed remarkable presence of mind: he told the pupils to evacuate the building without delay. He then went to fetch the school-keeper and the two of them returned just in time to see a roof beam collapse (see Plate 7.1). Some six hours later, while the debris was being examined and preparations were being made to prop the roof, a second beam, adjacent to the one that fell in the morning, collapsed (see Plate 7.2). With it, three bays of roof cladding came down.

A full report[35] on this occurrence and its causes was prepared by the Building Research Establishment and this is the source of much that follows.

Structure

The school was built in 1965 and 1966; the consulting engineer was the former chief engineer of Pierhead Ltd. who supplied the prestressed concrete beams. The two buildings of interest, the swimming pool hall and the gymnasium, are similar in construction. Their internal dimensions are: length 21.3 m; width 10.1 m in the case of the swimming pool and 10.7 m in the gymnasium; height of both buildings 4.9 m to 5.0 m, in order to provide a fall for roof drainage (see Fig. 7.1). The walls are built of load-bearing brickwork topped by a reinforced concrete ring beam cast in place using Portland cement. Into this ring beam, over a length of 100mm, there were cast the ends of precast prestressed concrete beams spaced at 1.5m.

Figure 7.1 **General arrangement of the swimming pool at the Stepney school[35].**

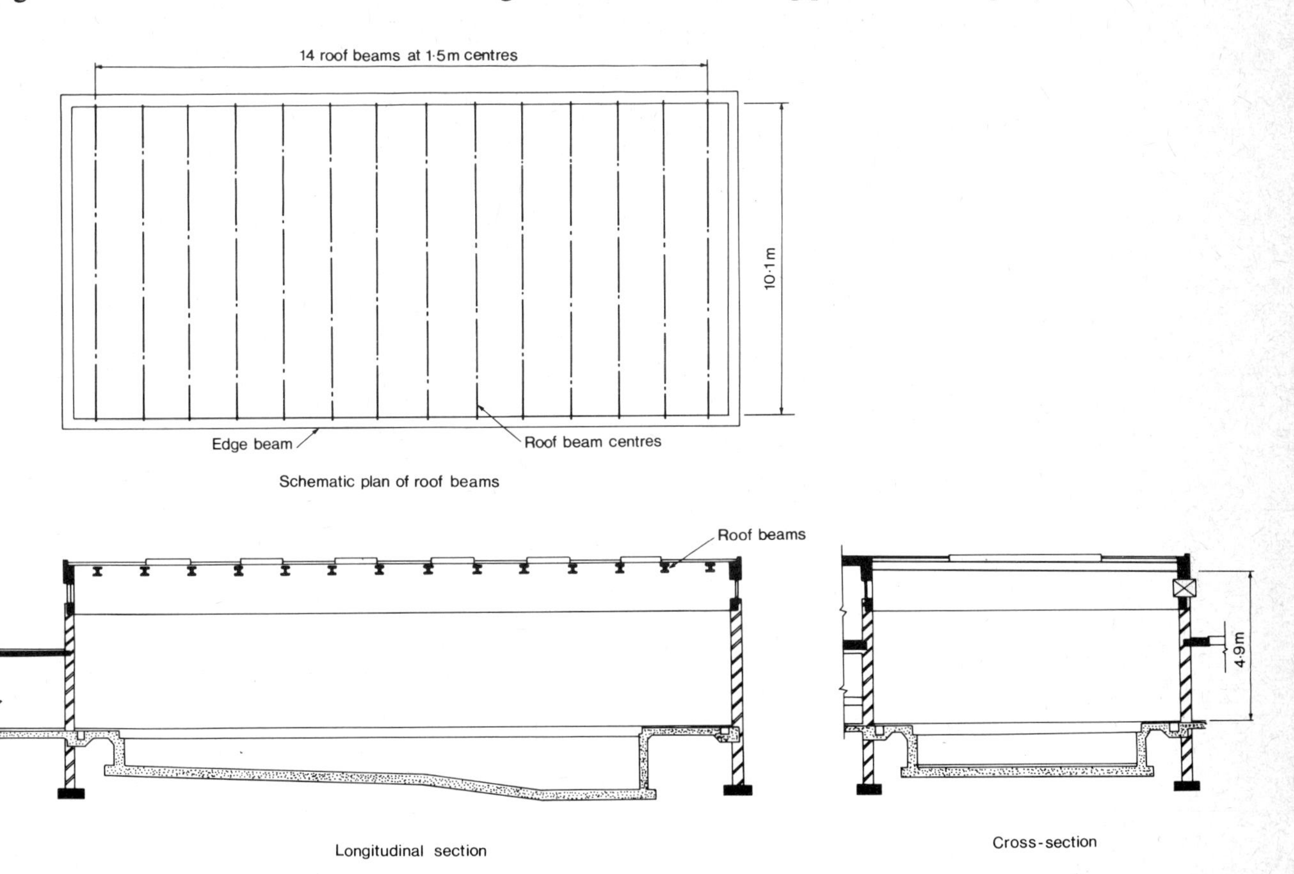

In the swimming pool building, these joists supported woodwool slabs 50 mm thick, on top of which there was a19 mm Portland cement and sand screed with a final finish of asphalt, 19 mm thick. The underside of the woodwool slabs was coated with 12 mm of plaster. Cast-in-place rooflights were supported by the joists.

The gymnasium roof was similar, except that there was no plaster coating on the soffit, and in the five central bays the joists were made to act compositely with a structural concrete deck in order to support additional load from climbing apparatus.

The roof beams were made of HAC concrete during the period 1st to 16th June, 1965; the temperature at the time is relevant to the discussion on the causes of the collapse and is referred to later in this chapter.

Previous inspection

The buildings appear to have been sound, at least insofar as nothing untoward was observed during re-decoration in 1970 and again in March 1973. Following the collapse at the Camden school (see Section 6.1), the roofs of the two buildings at Stepney were inspected by the designer. This took place in September 1973, and in October 1973 he *"concluded that there was no evidence of any damage or distress in the joists (beams), nor any undue deflection, to give rise to any cause for anxiety with regard to structural adequacy. The strength of the concrete in the joists (beams) had*

Figure 7.2 Cross-section of the roof beams at the Stepney school[35].

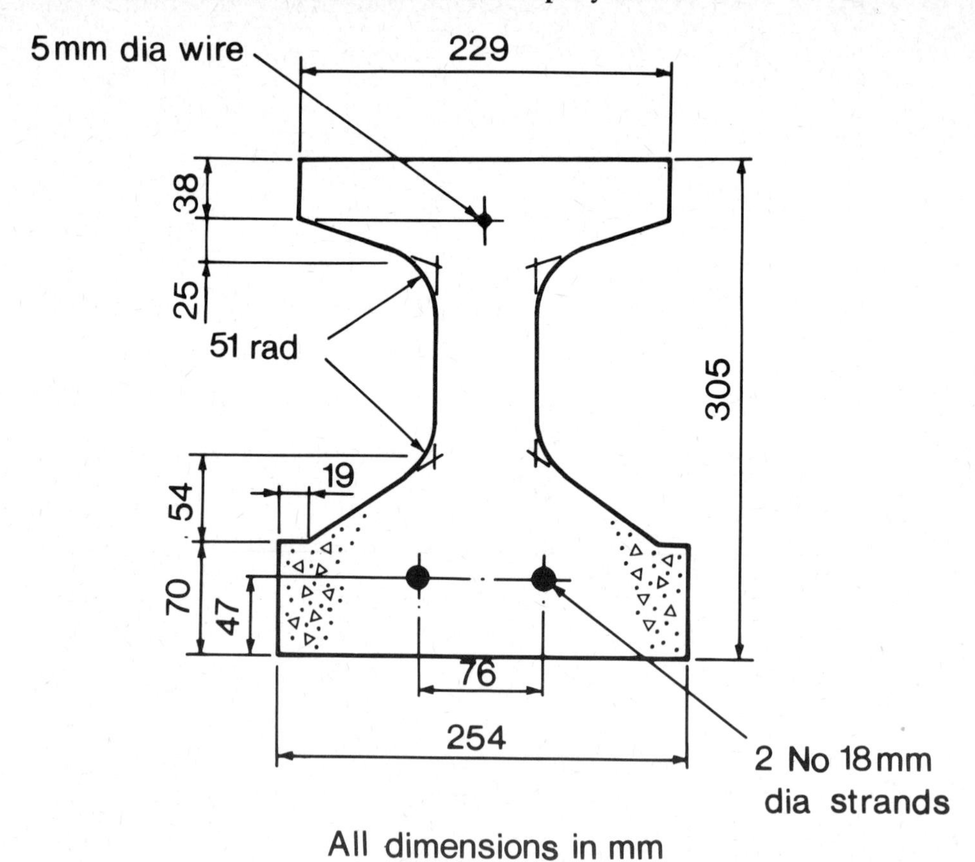

been checked in his survey with rebound hammer equipment which had indicated a compressive strength of 52 to 83 N/mm²"[35].

There are two important aspects of the consulting engineer's report. First, it seems that it is not easy to inspect a building and predict its performance even in the near future. (The collapse took place less than five months after the inspection.) Secondly, the use of a rebound hammer in assessing the soundness of HAC concrete is even less reliable than in the case of Portland cement concrete; indeed, the hammer is of no use.

Examination after collapse

The examination of the structure following the collapse was hampered somewhat by the fact that, two days after the beams had fallen, the whole roof of the swimming pool was removed in the interests of safety. It seems, however, that there was no untoward deflection in the beams that did not collapse. The removal of the beams was not easy. It is reported[35] that "*it had been found much easier in demolition to cut through the HAC concrete of the roof beams than through the Portland cement concrete of the edge beam*". It is important to note that "*The reinforcement and tendons in the prestressed concrete beams showed no signs of undue corrosion*"[35].

As far as the concrete in the beams is concerned, the description in the report[35] of the Building Research Establishment is of interest: "*The site examination of the debris showed that the concrete in all the beams from the swimming pool was very friable and crumbled easily, particularly when it was wet. The concrete was brown in colour, the fractured surfaces revealed a preponderance of aggregate particles of about 18 mm in size, suggesting that a gap-graded mix was used, and adhesion between the mortar and this aggregate was weak. It was particularly noticeable that most of the surface*

Beam No. †	No. of cores tested	*Ultrasonic pulse velocity (km/s)* Maximum	Minimum	Mean	Standard deviation	*Equivalent cube strength (N/mm^2)* Maximum	Minimum	Mean	Standard deviation
S3	3	4.00	3.00	3.50	–	15.6	12.4	13.8	–
S5	2*	3.56	3.49	3.52	–	25.4	21.4	23.4	–
	6	4.00	3.55	3.70	–	18.0	12.7	16.1	–
S6	5*	3.41	3.02	3.25	–	21.3	13.7	16.3	–
	5	3.50	3.23	3.40	–	16.8	14.1	15.7	–
S10	5*	4.10	3.70	3.90	–	22.9	14.6	19.8	–
	4	4.05	3.33	3.65	–	27.7	18.3	22.7	–
S11	20	4.36	3.70	4.10	0.16	26.6	19.5	22.2	2.0
S12	3	4.20	3.75	4.05	–	22.5	19.4	21.3	–
S13	6*	4.00	3.26	3.70	–	19.7	18.7	19.1	–
	8	4.00	3.43	3.75	–	19.4	15.9	17.6	–
S14	15	4.30	3.61	3.95	0.19	30.5	22.1	25.5	2.5
G2	9	3.90	3.05	3.60	–	23.6	14.4	17.9	–
G4	12	4.15	2.75	3.50	0.38	21.3	17.0	18.4	1.3

* Cores tested dry   † S = swimming pool roof   G = gymnasium roof

Table 7.1   **Results of tests on cores – ultrasonic pulse velocities and equivalent cube strengths for the Stepney School[35]**

of the prestressed concrete beams consisted of a very hard skin of about 3 mm thickness, which was grey in colour and was composed of cement and fine aggregate only. It was also noticeable that some of the fractured surfaces of the two beams that collapsed revealed the presence of a white crystalline material, which was later identified as ettringite. Crystals of calcite were also visible in many of the beams."

It is the hard skin that probably explains why the rebound hammer gave such a misleading picture of the strength of the concrete in the beams.

Strength and ultrasonic pulse velocity

Since the actual values of strength and of ultrasonic pulse velocity are of interest, we are reproducing the data from the Building Research Station report[35] in Tables 7.1 and 7.2. The ultrasonic pulse velocities were determined across the top flange of the beams and the cores were cut vertically down through the flange; they were 75 mm in diameter and 75 mm long. The report[35] notes that in some instances, mostly in the gymnasium beams, *"there was an absence of coarse aggregate from quite large areas in the top flanges of the beams"*. The strengths reported in Table 7.2 are those of equivalent cubes, i.e. the core strengths, corrected for the length-diameter ratio, were converted to cube strengths using the British Standard BS 1881:Part 4:1970. Table 7.1 shows that the equivalent cube strengths ranged from 12.4 to 30.5 N/mm^2 for the swimming pool roof beams and from 14.4 to 23.6 N/mm^2 for the gymnasium roof beams. We may view these values against the minimum • strength for pretensioned concrete laid down by the Code of Practice for Structural Use of Concrete CP 110:1972, viz. 40 N/mm^2. (Admittedly, the design anteceded the code.)

The measured values of strength should be compared with those determined at the time of manufacture of the beams.According to the report[35], *"The average strength of two cubes, tested at one day and cast from batches mixed on the same day as the beams were cast, ranged from 54 to 75 N/mm^2 with a mean value of 63 N/mm^2 and a standard deviation of 6.6 N/mm^2. The mean strength for the beams in the swimming pool roof was 58 N/mm^2 and in the gymnasium roof was 68 N/mm^2. The free water-cement ratio for these batches, determined by siphon-can test, varied between 0.36 and 0.40. The modulus of elasticity of the concrete, as determined from tests on smaller joists cast on the same days, was 38 kN/mm^2."*

Mix proportions

The mix used was apparently 51 kg of HAC, 154 kg of sand with 6 per cent moisture, and 295 kg of coarse aggregate with 1 per cent moisture[35]. The siphon-can test was used to measure the moisture content of the concrete and this test is of doubtful accuracy. Presumably, in reality the water content was controlled by workability. Indeed, tests on the concrete from the beams after collapse led the Building Research Establishment to the conclusion[35] that the probable (indicative) value of *"the free water-cement ratio was in the range of 0.45 to 0.55 with a most likely value of 0.48, but the estimate depends on assumptions that are not directly measurable"*. From the cement suppliers' tests on concrete made for checking the quality of the cement delivered for the manufacture of the actual beams used at Stepney, one would expect higher strengths than were actually obtained by the precast concrete manufacturers, had the water-cement ratio really been between 0.36 and 0.40. The assessment[35] by the Building Research Establishment of these strength values obtained by the concrete manufacturer is *"that a substantial amount of the concrete tested had a free water-cement ratio of more than 0.4"*.

Table 7.2 **Ultrasonic pulse velocities at different positions determined in different conditions with corresponding strengths for two beams in the gymnasium roof of the Stepney school[35]**

Beam No.	*Distance from the end of the beam m*	*Ultrasonic pulse velocity (km/s)* Across top flange *in the roof**	*in the laboratory*	*Core*	*Equivalent cube strength of core N/mm²*
G2	0.57	4.10	3.85	3.50	14.9
	0.92	4.15	3.85	3.55	17.9
	1.16	4.10	3.70	3.70	15.5
	1.54	4.05	3.75	3.90	23.6
	2.18	4.15	3.80	3.85	21.5
	2.43	3.95	3.05	3.75	16.9
	2.79	4.20	3.40	3.45	20.5
	3.03	4.00	3.70	3.55	15.9
	3.50	3.80	3.05	3.05	14.4
Mean		4.05	3.55	3.60	17.9
G4	0.49	3.90	3.80	3.85	17.1
	0.90	4.00	3.95	3.30	17.0
	1.15	3.85	3.65	3.55	17.3
	1.52	4.00	3.95	3.60	19.0
	1.77	4.05	3.80	3.45	18.6
	2.16	4.10	3.45	2.75	18.4
	2.40	4.10	3.70	4.15	21.3
	2.78	3.90	3.65	4.00	20.3
	3.00	3.95	3.65	3.65	19.0
	3.36	3.90	3.80	3.45	17.8
	3.59	3.85	3.60	3.25	17.4
	4.01	3.80	3.45	3.20	17.5
Mean		3.95	3.70	3.50	18.4

* These velocities were measured close to but not necessarily precisely at the distances given which relate specifically to the laboratory tests and core positions

Deflection and cracking

An inspection[35] of the gymnasium roof showed that nearly all the beams had sagged. The measurements of the sag are given in Table 7.3. More information on the gymnasium roof is given in the report[35] : *"Some cracking was seen in both the edge beams and the roof beams. In the edge beams there were occasional diagonal cracks below and away from the edges of the bearings of the roof beams. Some vertical cracks were also visible near the ends of the prestressed roof beams where they were cast into the edge beams. Neither of these occurrences of cracking were regarded as structurally significant, but it seems likely that similar cracking occurred in the swimming pool roof. Since some of the gymnasium roof beams had sagged appreciably, as much as 34 mm, ultrasonic pulse velocities were measured at fairly close intervals across the top flange of a number of the beams to obtain a relative indication of the quality of the concrete. The larger amounts of sag were associated with the lower pulse velocities which indicated that the strength of the concrete was suspect."*

Table 7.3

Measured sag and ultrasonic pulse velocities for beams in the gymnasium roof at the Stepney school[35]

Beam No.	*Measured sag at mid-span mm**	*Ultrasonic pulse velocity (km/s)*	
		Mean	*Standard deviation*
G 1	21	4.25	0.12
G 2	34	4.00	0.17
G 3	15	4.10	0.14
G 4	31	4.00	0.15
G 5†	9	–	–
G 6	6	–	–
G 7	12	4.50	0.11
G 8	6	–	–
G 9	9	–	–
G10	9	–	–
G11	6	–	–
G12	0	4.20	0.15
G13	9	–	–
G14	24	4.00	0.20

* From information provided by Messrs. R.J. Crocker and Partners

† Composite beam. Beams G5 to G10 were constructed compositely with the roof.

Beam type

The beams used at Stepney were of the type shown in Fig. 7.2, and known as X12 joists. They were originally developed for use in composite construction with a cast in-situ concrete topping but have also been used for light roof construction with non-composite cladding. The design ultimate strength of the beams as calculated for the specified concrete cube strength of 62 N/mm^2 at the age of 28 days was 108 kNm, using a method of calculation based on the Code of Practice CP 116:Part 1:1965. Some of the beams, or their portions, removed from the two roofs were tested to destruction and developed ultimate moments shown in Table 7.4.

Analysis of cores showed a high degree of conversion both in the swimming pool and in the gymnasium roof beams. This applies to all the samples taken with the exception of the surface skin. Once again then we see that surface assessment of the 'state of health' of HAC concrete is unreliable. The actual values in the swimming pool roof ranged between 75 and 95 per cent, with 86 per cent as the modal value[35]. For practical purposes, this is virtually a state of full conversion. Of the gymnasium roof only two beams were tested. The degree of conversion ranged from 50 to 85 per cent.

The white crystalline material referred to earlier (page104), identified as ettringite, *"was the result of sulphate attack, the sulphate being probably derived from the gypsum plaster applied to the under surface of the woodwool*

Beam No.†	Test arrangement	Behaviour in elastic range of loading: Depth of neutral axis for bending mm	Modulus of elasticity of concrete (kN/mm^2) from deflection	from curvature	from strains	Total moment at failure kNm	Mode of failure
S 1	Loading at two points 1.22m apart symmetrically spaced on a span of 6.10m	178(a)	29	–	29 to 34	47	Inclined to shear with cracking under end load point (c)
S12	Loading at two points 1.52m apart symmetrically spaced on a span of 7.01m. Ends encased in reinforced concrete	187(a)	–	33	34 to 50	70	Flexure
S14	Third point loading on a span of 4.65m	198(b)	25	–	16 to 20	48	Inclined shear (c)
G 2	Four-point loading of beam in position in gymnasium roof	169(b)	45	–	32 to 49	84(d)	Flexure
G 2 G 4	Loading at two points 1.52m apart symmetrically spaced on a span of 4.11m	200(a)	23	28	28 to 29	–	Stiffness test only
G 4	Four-point loading of beam in position	180(b)	48	–	37 to 52	85(e)	Flexure
G 4	Loading at two points 1.52m apart symmetrically spaced on a span of 4.57m	193(b)	29	30	31	–	Stiffness test only
G 4	Loading at two points 1.52m apart symmetrically spaced on a span of 4.88m. Ends encased in reinforced concrete.	185(a)	–	31	29 to 33	91	Flexure

Notes: (a) mean for three positions; (b) for one position only; (c) failure influenced by splitting along tendons;
(d) the additional deflection immediately before failure was 40mm; (e) the additional deflection before failure was 25mm;

† S = swimming pool roof; G = gymnasium roof

Table 7.4 **Summary of results of flexural tests on beams for the Stepney school[35]**

slabs[35]." It was found that where the sulphate attack had occurred there was no hard skin on the concrete.

The report[35] states further: *"No signs were found of alkali hydrolysis, i.e. attack by alkali from the Portland cement in the screed."* Comments on the use of feldspars and micas in both the coarse and fine aggregate used were quoted on page 66. The use of aggregate with feldspars and mica in 1965 may be viewed against the background of our warning in 1963[45] and of the report of the Institution of Structural Engineers in 1964[76].

Beam manufacture

In the course of the enquiries by the Building Research Establishment it was found that the roof beams in question were not made under cover but in the open. The manufacturer supplied information[35] on the weather conditions during the casting period, as follows: *"The maximum temperature was 24°C with a range from 17°C to 24°C for the maximum for each day; on 6 days there was more than 10h of sunshine and, while there was generally little or no rain, 6mm or more was recorded on 2 days. No information is now available on the conditions introduced for curing or later storage."*

It is hard to imagine that the manufacturer was not aware of the fact that hot weather during casting and curing prior to transfer is distinctly harmful to HAC concrete. With all the warnings about casting temperature, why did they work in the open in the summer?

Causes of failure

Let us now look at the causes of failure as established by the Building Research Establishment[35] : *"The failure of the two beams in the roof of the swimming pool building is attributed to a combination of two factors. Firstly the high alumina cement concrete had converted considerably with a substantial loss of strength and, as a consequence its resistance to chemical attack was reduced; it was then attacked by sulphate which led to its disruption.*

Gypsum plaster, which had been applied to the woodwool slabs between the beams, was in contact with the concrete and was therefore the most likely source of sulphate. The presence of free water is necessary for the attack to take place. There was evidence of substantial condensation at roof level and leakage of rain water through the rooflights may well have occurred; either source may have provided the necessary wet conditions. Disruption of the converted concrete then occurred as a result of the growth of ettringite crystals within the concrete itself."

And further[35] : *"The average temperature in the roof beams over the swimming pool was assessed as being between 27°C and 28°C, and since the use of high alumina cement concrete is discouraged at temperatures above 27°C, as in CP 116:Part 1:1965 for example, conversion with some loss of strength was expected. The extent of the loss and the degree of conversion was, however, unexpected for concrete of the specified quality and age. The degree of conversion and the loss of strength for the concrete in the gymnasium roof, where the average temperature of the beams was 17°C to 18°C, was totally unexpected. Since the conditions in this building were no more onerous than for most buildings using this type of construction with high alumina cement concrete, increased emphasis was given to the need to examine the quality of the concrete used in the beams, the possible temperature conditions during their life that might have affected their strength, and the nature of the aggregates ."*

These are important words but the finding is probably not unexpected to those who are familiar with the literature on HAC reviewed in Chapters 3, 4 and 5.

Conversion

Having established the presence and total degree of conversion, we may want to know the degree of conversion soon after casting. Unfortunately, it is rarely possible to draw a demarcation line between early and late conversion (cf. Chapter 5). For the user of the building, this may be of no more than academic interest but, if "blame" can be attributed to the early or the late conversion, this may be of significance in determining responsibility for the occurrence. The report[35] says: *"The curing conditions for the beams cannot be defined precisely. Since the beams were cast in the open and may not have been covered during their early life it is possible that, with shade temperatures reaching a maximum of about 20°C and with about 10h of sunshine, the temperature of the dark horizontal surfaces could have reached 38°C and been in excess of 35°C for 2 or 3h during the early afternoon. If this peak coincided with the development of heat due to hydration of the cement, the internal temperature might have risen by a further 4°C or 5°C. Thus there is a possibility that some of the deterioration in the strength of the concrete was due to conversion at a relatively high temperature whilst the beams were on the prestressing beds."* And further: *"A high temperature during the first few hours after casting the beams would also explain why the surface skin of the concrete became dehydrated and therefore converted to only a small degree without changing colour and without apparent loss of strength."*

Conclusions

Having read all this we may ask the question: should these beams and this design have been used for the Stepney buildings? The report[35] of the Building Research Establishment says: *"The design of precast prestressed concrete members using high alumina cement is covered by the Code of Practice CP 116:Part 1:1965. It recommends permissible stresses and design strengths for concrete for two specific classes of service. Class I covers those conditions where the water-cement ratio is not greater than 0.4, and the concrete is kept cool and moist during the first 24h, and reasonably dry and cool after its initial curing. The design of both roofs did not depart significantly from these recommendations for strength and permissible stresses given for Class I service conditions. As already discussed, however, it is doubtful whether the control of water-cement ratio and of curing met the requirements, and for the swimming pool the additional requirement for cool conditions in service was not met. The permissible stresses and design strengths for concrete appropriate to Class II service conditions are such that it would be impracticable to use high alumina cement concrete in this form of beam since the assumed strength of concrete for design is reduced from 52 N/mm² (7500 lb/in²) for Class I service conditions to 22 N/mm² (3250 lb/in²)."* (see page 116).

The report then continues to say that, although the tests carried out in the development of the X12 joist (beam) indicated an adequate load-carrying capacity, *"the top flange of these precast concrete beams was of appreciably smaller size than the bottom flange and it was well suited to use in composite construction where the compressive stresses in flexure are mainly resisted by a structural topping. For such construction, a loss of compressive strength in the precast unit is not of great significance so long as the topping retains its strength. If, however, the precast units of this shape are used without composite action, any deterioration in the compressive strength of the concrete leads to a direct reduction in the strength of the member. The X12 joist is not, therefore, well suited for use in individual members made of high alumina cement concrete."*

The final conclusion in the report[35] is that *"the cause of failure of the roof beams over the swimming pool . . . was loss of strength due to conversion of high alumina cement concrete followed by chemical attack leading to disruption of the concrete. The chemical attack, which was localised, was probably due to the action of sulphate, derived from the gypsum plaster in contact with the beams, taking place in the presence of water from condensation or leakage at the rooflights*." It is also stated that *"Further investigation led to the inferences that:*

(i) the free water-cement ratio may have exceeded the maximum of 0.4 aimed at in production;
(ii) the temperatures during the first day after casting may have been excessive; and
(iii) the aggregate used which contained the minerals feldspar and mica had an accelerating effect on the rate of conversion and loss of strength."

We would like to make an additional comment. The report[35] of the Building Research Establishment refers to an unexpected degree of conversion in the roof beams of the gymnasium and to the unexpected loss of strength in the beams over the swimming pool. We feel that these are really not unexpected in view of the discussion in Chapters 3, 4 and 5. Be that as it may, from now on we should be prepared for losses of this magnitude in HAC concrete beams in our buildings.

As a final remark, it may be interesting to mention that during the 1974 summer school holidays both the swimming pool and the gymnasium roofs at the Stepney school were re-designed and replaced; in fact, they were ready when the school re-opened on 4th September, 1974. This was a piece of rapid work. Steel beams with precast concrete deck units (made with Portland cement) were used. The cost of the repair was reported in *Building* of 27th September, 1974[109] to be in excess of £50,000.

8. Rules on the Use of High-Alumina Cement in Structures

8.1 Situation abroad

France

As a result of the various failures, many of which have received no publicity, restrictions on the use of HAC have been imposed in different countries. France, where HAC was first used, was also first to ban its structural use. This came about as follows.

Because of the rise in temperature during placing, in 1928, the French Ministry of Public Works[77] restricted the cement content (when HAC is used) to a *maximum* of 400 kg/m^3 of concrete. Since, however, such concrete tended to be insufficiently impermeable, especially in marine structures, in 1935 the Ministry[78] decided that 400 kg/m^3 of concrete should constitute a *minimum* in structures in which impermeability is of essence. In cases where HAC is used, special precautions to avoid a high rise in temperature were specified. Failures, some of which are described in Table 5.1, led the Ministry of Public Works, in accordance with the views of the Conseil Général des Ponts et Chaussées, to issue a new instruction in 1943[79]. This forbade the use of HAC in mortar and concrete in all permanent public works: highways, internal waterways, and marine works.

As an exception to the instruction, HAC could be used with permission of the Ministry of Works when the special properties of the cement made its use advisable but such a permission was granted only after satisfactory long-term tests. The cement could also be used in repair work, provided easy inspection was possible, and safety was not affected. HAC may, of course, always be used in the construction of temporary works, where its very rapid hardening makes it invaluable.

It should be stressed that there has never been any limitation on the use of HAC in private works in France. HAC has also been used there in jointing of precast beam segments to be post-tensioned and in grouting of ducts in such beams. The cement was also used successfully in urgent repair work of slabs at Orly airport.

The situation changed in March 1970 and a new circular was issued by the French Ministry of Construction[80]. The reasons given for this reversal are research by the cement makers and by several laboratories, but it is stressed that the new circular applies only to *ciment alumineux fondu* and not to HAC in general. This *ciment alumineux fondu* is defined as cement obtained by grinding, after burning to the point of fusion, of a mixture consisting mainly of alumina, lime, silica and iron oxide in the form of bauxite and limestone. The proportions have to be such that the resulting cement contains at least 30 per cent by weight of alumina and no more than 0.1 per cent of sulphur nor more than 0.4 per cent of alkalis (K_2O and Na_2O). We can note that British-made cement satisfies these requirements.

The circular[80] then lays down the conditions under which HAC, as specified, can be used without risk of later trouble. The first condition concerns the choice of aggregate. This must not contain sand with freeable alkalis, including granitic, schistose or micaceous material or any constituents in a more or less advanced state of decomposition, nor must it contain slag. Since alkalis are present mainly in the finest materials the minimum particle size is laid down as 0.2 mm; it is pointed out that the absence of the finest fraction in the aggregate is conducive to the use of a reduced water content in the concrete.

The second condition concerns mix proportions. The maximum permitted water-cement ratio is 0.40. The minimum cement content is laid down as 400 kg/m^3. No admixtures are allowed. Loss of cement paste through leaky formwork must be avoided.

The third item dealt with by the circular[80] is the control of temperature. When HAC concrete is not massive, surface drying may be prevented by spraying with cold water during a period of 48 hours from the time of removal of formwork. This should be done as soon as possible, and not later than four hours after placing in the case of non-load bearing members. In the case of mass concrete, a cooling system must be used.

The circular[80] stresses that the use of HAC is a *"delicate"* operation and points out the very considerable difference between the use of HAC and of Portland cement. With the latter, departure from specification does not as a rule lead to disaster, but in the case of HAC this is far from being the case. The minister under whose name the circular appears then states in italics: *"this is why I want to stress that the conditions listed above must not be deviated from."* In other words, the maximum water-cement ratio, the minimum cement content etc. must be applied rigorously. We cannot help thinking that if this had been done in the case of the structures discussed earlier in the book, some of the failures would have been avoided. (For instance, the cement content in the Stepney school beams was 240 kg/m^3.)

One final comment about the situation in France should be made: HAC is very little used structurally in France and, unlike the United Kingdom, there has never been a French precast prestressed concrete industry based on HAC.

Germany

Let us now move to Germany. The failures there were described in Section 5.4. In consequence, starting in 1962[81], the various *Länder* governments prohibited the use of HAC in the construction and repair of load-bearing structures made of concrete, plain, reinforced, or prestressed. The cement makers' proposal to permit a general use of HAC, with a description of the range of applications, was rejected. On the federal scene, all reference to HAC in the German codes for reinforced and prestressed concrete was withdrawn in 1962. This withdrawal applies also to chimneys for acid gases but the use of HAC in furnace construction is not affected.

It is interesting to note that a partial restriction on the use of HAC was introduced in Germany as early as September 1960, i.e. before the Bavarian failures of 1961. The effect of the restriction was to permit the use of HAC with a special permission only and using authorised procedures[69]. The reason for this was not corrosion but *"uncertain strength behaviour"*[69].

Scandinavia

Even a country as far removed from the tropics as Norway decided not to permit the use of HAC in building construction. The mention of this cement was deliberately[82] omitted from the February 1962 revision of the Norwegian Standard NS427A, although previously the use of HAC was permitted. Nowadays, HAC is virtually unknown in construction in Norway. Likewise, in Sweden, contrary to the previous standards, the 1960 standard for cement (B1) prohibited the use of HAC in structures[54]. In Finland, because of the negative experience with HAC there as well as in Sweden, the use of HAC in structures was forbidden nearly thirty years ago. The only current uses are in minor repairs and for fire-resisting purposes[83]. In Denmark, the standard for concrete DS 411 provides for the use of Portland cement only. Thus, the use of HAC in structural work is forbidden because *"the risk of loss of strength due to conversion outweighs the otherwise excellent properties of this cement."*[121] The cement is, however, used in construction and repair of kiln linings in tile- and brickworks, in foundries, gas works, and factory chimneys, in masonry repair work, and for acid-resistant floors. This is of course the case in other countries as well. In Hungary structural use of HAC was forbidden nearly 20 years ago.

Other countries

In some other countries no legal restrictions on the use of HAC are in force, but because of past adverse experience the use of this cement is extremely rare. This is the case in New Zealand, where, since the failure of large prestressed concrete beams in a power project construction, HAC has been used mainly in the manufacture of precast pig troughs and fence posts, and for other non-structural purposes. Likewise, in Australia, HAC is used mainly in repair work and for refractory purposes, and only very exceptionally in structures.

HAC is not mentioned in the Australian code for concrete structures, as 1480–1974, but cements other than Portland or blended can be used *"if approved and if the resulting concrete has strength, durability and characteristics not inferior to those required by this code for Portland cement concrete".*

In Spain, HAC used to be used extensively in the manufacture of small prestressed precast concrete beams, mainly for agricultural buildings. A number of failures occurred, none of them disastrous, and the structural use of HAC was discontinued. Formally, use in reinforced concrete is restricted and in prestressed concrete will be forbidden in the 1975 Code of Practice[129].

In Japan, there is no code provision for the use of HAC and it has never been used in structures to any extent owing to difficulties experienced in practice many years ago[130].

The European Committee on Concrete, in its International Recommendations for the Design and Construction of Concrete Structures[85], published in 1970, states that HAC *"should only be used after special justification."* The document adds that *"Cements for use in aggressive conditions (sea water, selenitic waters, etc.) should be specially accepted for this use by a properly qualified authority . . ."* We should note this reference to a *"properly qualified authority"* and not to the cement makers.

From this brief survey therefore it can be seen that in most countries of the world high alumina cement concrete is not normally used in structural situations. It is only used to any significant extent in special cases such as foundations or repair work where its particular properties make its use advantageous.

8.2 Situation in the United Kingdom

Traditionally, there are no banned construction materials and no regulations with the force of law. Guidance is given through the medium of codes of practice. However, before these dealt with HAC in any detail, the Institution of Structural Engineers published (in 1964) a report on *The use of HAC in structural engineering*[76]. The committee which wrote the report included consulting engineers, contractors, a member of staff of the Building Research Station, an engineer who had done research for the Lafarge Aluminous Cement Company Ltd., and the Author of this book (who, by the time the report was drafted, had moved to Canada). As with all committees, the final report was a compromise, and probably no member of the committee was in wholehearted agreement with every part of the report.

The report[76] was cautious. In the introduction, it is stated that the recommendations in the report *"relate to HAC of British manufacture only."* There is also a curious statement, as follows: *"Many reports of failure or partial failure due to deterioration in strength and durability of concrete made with high-alumina cement in which the deterioration has been attributed to 'conversion' have been studied, including those quoted by Neville*[45]*. Unfortunately, in many cases there is no detailed information available regarding precautions taken at the time of mixing and curing to minimize conversion effects. It is therefore possible, at least in some cases, that the mix proportions and curing conditions were such as would aggravate unfavourable conversion effects."* What conclusions was the reader expected to draw from this?

The report[76] concludes with the advice: *"High-alumina-cement concrete can be used safely for load-bearing members of a structure provided that the precautions outlined in the report are taken.*

High-alumina-cement concrete should not be used in a load-bearing structure in tropical conditions or in industrial structures where it will be subjected to warm and moist conditions, as full conversion can occur if even well-cured concrete is subsequently maintained under water or in saturated conditions at temperature above 25°–30°C (77°–86°F)

High-alumina-cement concrete can also be used to advantage where stress is not critical, but even in such cases a low water-cement ratio is recommended together with good compaction."

The report[76] was withdrawn in December 1973, as some of the recommendations in it were no longer regarded as satisfactory, but possibly also partly because of the failures described in Chapter 6.

The report of the Institution of Structural Engineers[76] served as a basis for the section on HAC concrete in the Code of Practice on the Structural Use of Precast Concrete CP 116: Part 1: 1965[13]. This code was metricated in 1969, and several statements from that edition of the code are worth quoting.

Class of service	*Type of construction*	*Specified works cube strength at one day N/mm²*	*Assumed concrete strength for design N/mm²*	*Permissible stress (N/mm²)*					
				Compressive		*Principal tensile (prestressed only)*	*Shear (reinforced only)*	*Bond (reinforced only)*	
				Direct	*Bending*			*Average*	*Local*
I	Reinforced and prestressed	50	50	13.7	18.3	1.20	0.90	1.00	1.50
	Reinforced only	40	40	11.0	14.7	–	0.90	1.00	1.50
II	Reinforced and prestressed	50	22.5	6.2	8.2	0.50	0.75	0.83	1.25
	Reinforced only	40	12.5	3.4	4.5	–	0.40	0.50	0.70

Table 8.1 **Permissible stresses in HAC concrete according to CP 116 : Part 2 : 1969[13]**

First, the temperature above which 'trouble' may be expected is assumed to be 27°C to 30°C: the code says *"In general, high alumina cement concrete is not recommended for structural units which will be maintained in wet or humid conditions at temperatures above about 27°C, and it should not be used in such situations without prior consultation with the cement manufacturer.*

Irrespective of the temperature and humidity conditions, high alumina cement should always be used according to the cement manufacturer's recommendation."

And later on, *"The use of HAC* (in hot, wet conditions at temperatures above 30°C) *is not recommended without prior reference to the manufacturers."* We should note the exhortation to consult the manufacturer, to which reference is made in Section 5.4. The Code also says that *"the rate and extent of conversion are negligible for properly cured concrete exposed to normal temperature and humidity conditions. Moreover, conversion is not invariably accompanied by a fall in strength, as at slow rates the decrease in strength may be negligible."*

When conversion is expected, CP 116:Part 2:1969 recommends the use of what are called Class II stresses. These are given in Table 8.1 and are based on the data for the converted strength given in the code and reproduced here in Table 8.2 and Fig. 8.1. We now know that these data are rather optimistic. Indeed, the report [35] on the Stepney school says *"The loss of strength and degree of conversion of the concrete was, however, greater than expected for the quality of concrete specified and exposed to these conditions."* It is also said in the report [35] that *"Recent research at the Building Research Establishment has shown that short periods of storage at slightly increased temperatures during the first day after casting may have an important effect on the strength of HAC concrete."*

Full details are worth quoting: *"For concretes with a free water-cement ratio in the range 0.35 to 0.45 stored at 18°C in water for one year, the strength will increase above the strength at one day by between 20 and 30 per cent. If similar concretes at the age of 3 h are placed in water at 35°C for 3 h and are thereafter stored in water at 18°C, the strength at 24 h is reduced by up to 10 per cent, but at one year the strength is reduced by up to 20 per cent. Thus the effect of the 3 h storage at 35°C is to change a 20 to 30 per cent increase in strength at one year into a reduction in strength of up to 20 per cent. Tests on strength at an early age do not indicate the magnitude of the effect and therefore if a temperature rise of this kind did occur in production it would not be indicated by the modulus of elasticity found in proof tests on*

Table 8.2 **Effect of water-cement ratio on the compressive strength of HAC concrete according to CP 116 : Part 2 : 1969[13]**

Water-cement ratio	*Compressive strength (N/mm²)*		
	A 1 day at 19°C	*B* 6 days at 38°C	*B as percentage of A*
0.35	66	49	74
0.40	63	39	62
0.45	59	31	53
0.50	54	24	44
0.55	49	19	39
0.60	44	15	34

Figure 8.1 **Strength of HAC concrete as a function of water-cement ratio, according to Code of Practice CP 116: Part 2: 1969[13].**

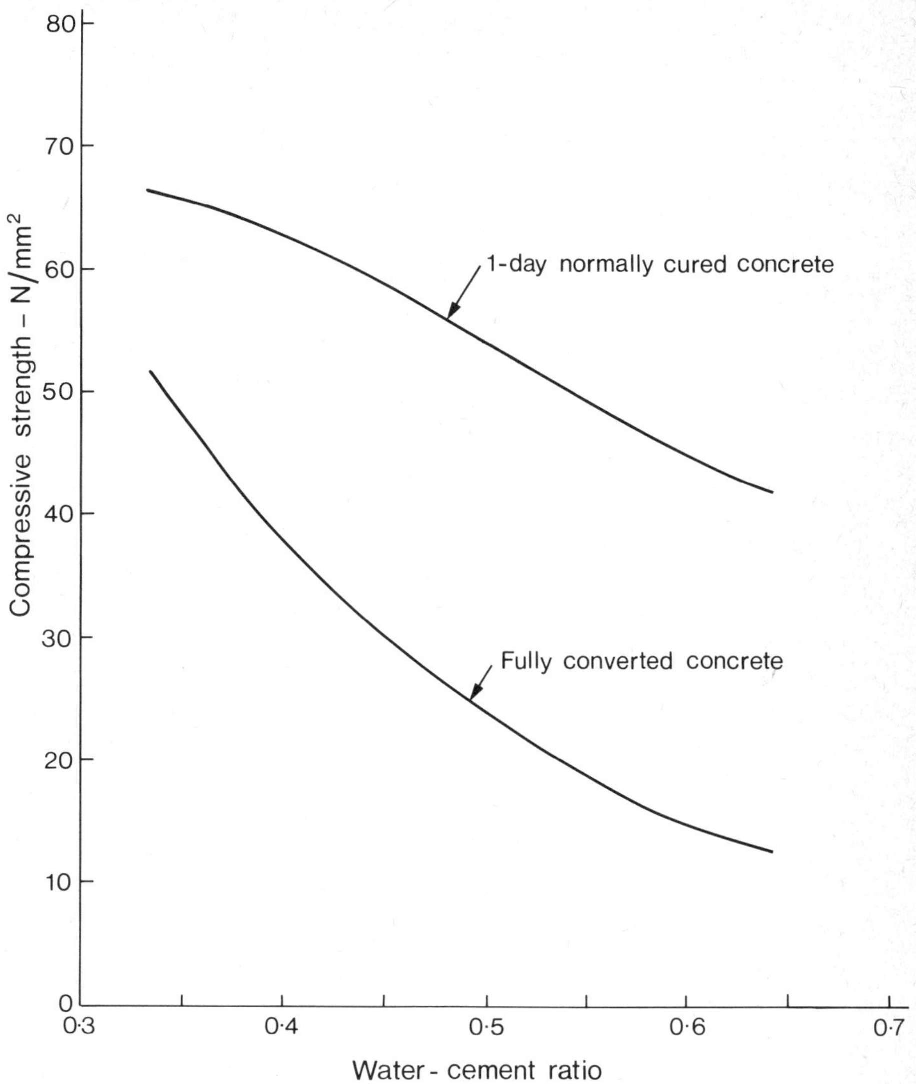

joists or by measurement of camber. If on the other hand, the period of 3 h had been spent at 25°C and not 35°C, the effect would have been quite different; the strength at one year was then found to be 20 to 40 per cent greater than the one-day strength of concrete stored continuously at 18°C. When similar concretes were placed in water at 35°C for 3 h immediately after casting without the delay period of 3 h, the effect of the temperature rise on strength at one year by contrast was not important. Since these tests were not continued for a period longer than one year, the effect of these treatments on minimum strength cannot be determined. It may be inferred however that, since the adverse effects on strength were increasing with time, the minimum strength would also be affected adversely. The sensitivity of high alumina cement concrete to relatively small differences in the temperature history of the first day after casting is however confirmed by these data."

Codes of Practice

Let us now revert to the codes of practice. The Code of Practice for the Structural Use of Prestressed Concrete in Building CP 115:Part2:1969[99],first published in 1959, recognizes as standard cements only Portland cement and Portland-blastfurnace cement. With respect to HAC, the code says that it *"should only be used with the Engineer's approval"*. We should note that the code covers both work carried out on site and the manufacture of precast prestressed concrete units. For precast work, however, the Code of Practice for the Structural use of Precast Concrete CP 116:Part 2:1969[13] seems to be at variance with the code CP 115 and allows the use of HAC as a standard cement. The code CP 116 says that the recommendations in CP 115, *"where applicable to the standard of structural precast concrete envisaged, have been reviewed and brought together in this Code* (CP 116) *with additional information and recommendations not previously published"*. The code CP 116 has already been considered in this section.

In December 1972, there was published for the first time a unified code[86] for the structural use of concrete of all types. Here was included a section on HAC concrete. Some of the provisions may be of interest and are discussed below. However, consequent upon the Stepney failure, the section on HAC concrete and all references to HAC were withdrawn in August 1974. The actual addition to the foreword to the code says: *"Recent research following structural collapses in 1973 and 1974 has considerably altered the understanding of the behaviour and sensitivity of high alumina cement concrete. As a result of both the research and the events themselves there is not, at this time, the consensus of opinion necessary for agreement on recommendations for the structural use of this material. Section 12, High Alumina Cement Concrete, of the 1972 issue of this Code has therefore been withdrawn and the Code no longer covers the structural use of concrete made with high alumina cement. Consequently an engineer wishing to specify concrete made with high alumina cement must rely entirely on his own engineering judgement without guidance from this Code of Practice."*

Thus HAC cannot be used if the design is claimed to comply with the code. This does not mean that a designer cannot use HAC: he can use any new or unusual construction material but he has to persuade the local authority of the soundness of his proposal. In the present climate of HAC problems with which many authorities are faced, this would be a Herculean task.

Having said this, we should nevertheless look at the main provisions relative to HAC in the 1972 Code of Practice for the Structural Use of Concrete[86]. The characteristic strength, which of course is well below the mean strength, is specified as 50 N/mm^2 and the maximum free water-cement ratio as 0.40. We are told that *"High alumina cement concrete should not be used in structural concrete which will be maintained in wet or humid conditions at temperatures above 27°C, unless designed for strength after full conversion.*

Where it may be difficult to prevent overheating during hardening, e.g. in members over 300 mm thick, or in fairly large sections cast in conditions which prevent water cooling of the concrete, or where the concrete may subsequently be used in wet or humid conditions at temperatures above 27°C, the structural design should be based on the fully converted strength as given in Table 62. The cement content should be such that the concrete has sufficient workability to permit thorough compaction at the chosen free water-cement ratio." This table is reproduced here as Table 8.3.

We are told to reduce the harmful effects of conversion, in addition to limiting strictly the water-cement ratio, by adopting the following measures:

"(2) By keeping the concrete cool and moist during the first 24 hours to ensure proper curing and to assist the dissipation of heat due to hydration

Table 8.3

Apparent characteristic strength of converted HAC concretes according to CP 110 : 1972[86]

Type of construction	*Characteristic strength* (N/mm^2)	
	Unconverted concrete (at one day)	*Fully converted concrete (apparent strength)*
Reinforced or prestressed	70	45
Reinforced or prestressed	60	30
Reinforced only	50	20

which is so detrimental at early ages. The following measures should therefore be taken.

a. *Wooden moulds (other than base or soffit forms) should be stripped or eased away as early as possible after casting (usually within four hours and not later than six hours). All exposed concrete surfaces should then be kept continuously wet by spraying with cold water for at least 24 hours. Further curing is then not normally necessary. Steel moulds need not be removed before applying the water.*

b. *When cooling is required, the use of wet hessian, a membrane curing compound, polyethylene sheets or other dry covering is no substitute for a water spray. Such materials should be used as the sole curing agents only for units in which there is little or no temperature rise (e.g. thin slabs, or thin-section beams in steel moulds).*

c. *In very hot weather the water for mixing and for curing should be as cool as possible (e.g. water which has been allowed to stand for long periods in the sun should not be used); aggregates should be shielded from the sun or sprayed with water before use. It is particularly important to strip the moulds and apply a water spray as early as possible.*

(3) By keeping the concrete reasonably dry and cool after its initial curing."

Provided these measures are fully observed, the design of reinforced and prestressed concrete members should be made in accordance with the parts of the code which apply also to Portland cement concrete (i.e. Sections 2,3,4 and 5), but *"Where adequate precautions are not taken to prevent conversion or where service conditions are such that full conversion is likely, the structure should be designed on the assumption that full conversion will occur. Sections 2,3,4 and 5 are still generally applicable but an apparent characteristic strength should be substituted for the characteristic strength,* f_{cu} *throughout. The apparent characteristic strength should have the appropriate value given in Table 62."*(Table 8.3 in this book).

Building regulations

Let us now look at the Building Regulations promulgated by the Department of the Environment. On 9th August, 1974, the Department[87] announced formally its intention* to amend the Building Regulations 1972 (as

*The original announcement was made on 20th July, 1974[92].

amended in 1973) so that local authorities would be able to reject plans for structural work involving the use of HAC. The announcement points out that *"The Department of the Environment have already recommended that high-alumina cement concrete should not be used for structural work until further notice . . ."* and says further: *"The possibility of prohibiting the use of high-alumina cement in all structural work has been considered, but it has been decided, on the basis of present information, to propose that no structural work using high-alumina cement shall be deemed to satisfy the mandatory requirements of Part D of the Building Regulations. This will enable it to be used in those few situations where its sulphate resisting and refractory properties make it the only suitable material to use."*

There is a further note[87] that the use of HAC in flue blocks and in dense concrete blocks in chimneys and the use of HAC in jointing and pointing are, because of the acid resistant and refractory properties of the cement, unaffected by the proposed changes to the Building Regulations. The possibility of use of HAC for the manufacture of ordinary building blocks, which is permitted by the British Standard BS 2028:1364:1968, is of no practical consequence since, according to the Concrete Block Association, HAC is not, in fact, used for the purpose[87].

The reasons for the proposed change in the Building Regulations are spelled out in some detail, it being said, among other things, *"that the factors contributing to loss of strength in high-alumina cement concrete are complex and it is possible that they have not all been identified. It is moreover apparent that high-alumina cement concrete is vulnerable to loss of strength from changes in temperature and humidity which may occur at any time in the life of the building, and that there is no way of accurately predicting the future extent of deterioration or when or over what period it is likely to occur."*

The amendment was laid before Parliament on 9th December, 1974, and came into operation on the 31st January, 1975. The amendment consists in essence of a statement that the use of HAC for structural work, including foundations, shall not be deemed-to-satisfy the requirements of the Building Regulations.

Prior to the amendment, at least two local authorities banned the use of HAC in structures. Croydon Borough Council did so as early as February 1974 both for public and private buildings (see Section 14.2). In June 1974, the City of Birmingham imposed a ban on HAC on the basis of regulation B1 of the Building Regulations, which refers to the use of suitable materials. Conceivably, this ban could have been challenged but, in the present climate of opinion, no one would assert that the ban discriminates against a good structural material. Indeed, the production of HAC concrete units by the main manufacturers ceased between May and July 1974. The Greater London Council is proposing to forbid the use of HAC *"in any building or part of a building without the approval in writing of the Council in each particular case"*.

There the matter rests: HAC is unlikely to be used in structures, including foundations, anywhere in the United Kingdom. Our position with regard to HAC is now no different from that in the rest of the world.

9. United Kingdom Government Circulars

The preceding chapter deals mainly with the situation prior to the 1973 and 1974 collapses. Let us now turn to the subsequent circulars of the various government departments.

June 1973

August 1973

Following the Camden school collapse, the Department of the Environment issued on 27th June, 1973 a circular[88] addressed to all the Local Authorities in England, instructing them to check their own buildings which are of the Camden form of construction. On 17th August, 1973 a further circular[89] was issued, in which it is stated: *"The form of construction potentially involving the weakness revealed by the Camden collapse has been used in both single and multi-storey buildings. It would typically comprise relatively long span precast concrete floor or roof beams or slabs supported at either ends on bearing nibs or ledges, and might be associated with either framed or load-bearing construction. The beams themselves will generally be hidden from view by a false ceiling.*

The technical examination of such buildings would normally begin with an appraisal of the overall stability of the structure followed by a critical examination of the local stability of the bearings."

The causes of the Camden school failure are reviewed and it is said: *"The combination of structural weaknesses shown in the report is unlikely to be widespread but methodical checking under professional supervision should be undertaken as soon as possible."*

February 1974

After the Stepney collapse, on 28th February, 1974, the Department of the Environment[90], acting also on behalf of the Department of Education and Science, wrote to all the Local Authorities in England advising them of the collapse and its possible causes and saying: *"In the meantime, it is suggested that you should identify as far as possible any parts of your own schools, and of any other buildings of post-war construction, whose roofs are of a similar type to that in the Stepney school, particularly those which are likely to have high temperatures at roof level, e.g. swimming baths, assembly halls, laundries, domestic science rooms, kitchens or industrial buildings. We suggest that you should then arrange, giving priority to places where people congregate in large numbers, for an appraisal of the structural design, possibly in assocation with the manufacturers, who have expressed their readiness to co-operate. It is not possible at this stage to suggest any simple field test for detecting weakness of the kind which has become apparent at the Stepney school, but a description of the building and some notes by the*

Building Research Establishment on the assessment of structures of this type are given in the Annex. If an obvious risk is identified you will then no doubt wish to consider interim safety measures such as propping the beams, or possibly, closure of the affected part of the building until further advice can be given in the light of the results of the Building Research Establishment's investigation.

The Stepney building was not the same in structural form as the Camden School for Girls where a roof collapse occurred in June 1973 (DOE Circular 109/73 of 17 August 1973 refers). It would be wrong therefore to assume that inspection and testing for faults of a kind which emerged at Camden would necessarily rule out a possible failure such as at Stepney; and buildings which have been looked at following the Camden incident should not be left out of account on this occasion.

It is also suggested that if you are aware, from your own knowledge, or after consulting records, of privately owned buildings of the same type in your area, you should consider informing owners at once and suggesting to them that similar precautionary checks should be carried out."

Useful advice on the assessment of structures with precast prestressed beams of HAC concrete is given in the same letter [90]. We would like to quote some of it:

"(a) Composite action *If, as in the school at Stepney, there is no composite action between the beams and structural deck, screed or topping, there is an increased risk that loss of strength in the concrete of the beams will lead to collapse. The likelihood of collapse will depend on the degree of the loss of strength and the level of stress in the materials under service loading. If there is no composite action collapse may be sudden and this is more likely with beams of I section with thin flanges.*

(c) Ambient temperature of beams *Temperatures at ceiling level may be considerably higher than those assumed in design due to high level heating or lighting (whether these were an original feature of the building or not) or solar gain through glazing or uninsulated construction. Assessment of the temperatures in the beams in the assembly hall of the Camden School for Girls showed that they were higher than might have been expected partly because lighting was accommodated within the ceiling space. In that instance it was concluded that 'on a sunny summer day the temperature in the roof beams could very occasionally reach 25°C . . . with the lights on for several hours; however, temperatures could rise to 25°C fairly frequently and occasionally to 30°C with average winter conditions'. The assessment of temperature in the roof beams is therefore an essential part of the appraisal of their strength and thus of the risk of collapse.*

(e) Water-cement ratio *The water-cement ratio used in high alumina cement concrete mix cannot be determined from the hardened concrete. Where information on free water-cement ratio is not available, it will be necessary to make the appraisal taking account of the fact that when high alumina cement was first used in prestressed concrete beams, the water-cement ratio was usually about 0.60. During the past 20 years, the importance of using a low water content has been increasingly recognised so that beams made 10–20 years ago might be assumed to have a free water-cement ratio of 0.5, unless otherwise stated. More recently it has become common practice to require the free water-cement ratio to be no more than 0.4.*

(f) Chemical attack *If water is in contact with the concrete as a result of leakage or condensation, such water may leach chemicals capable of attacking high alumina cement concrete from adjacent materials in the structure. The incidence of attack and its effect on strength is likely to be greater in concretes having lower minimum strengths on conversion.*

(g) Deflection *Deflection of beams below the horizontal may indicate loss of strength although an upward deflection does not necessarily imply that strength is adequate.*

(h) Test methods *Ultrasonic measurements of pulse velocity are likely to indicate the presence of seriously weakened concrete. The use of rebound hammers on the surface of the concrete, for example the Schmidt Hammer Test, may give misleading results since there is evidence of the occurrence of hard material on the surfaces of some weak concretes.*

In certain cases, it may be appropriate to carry out loading tests but special experience is needed in the methods employed and in the interpretation of the results."

May 1974

A further circular[91] to all the Local Authorities in England was issued by the Department of the Environment, again on its own behalf and on that of the Department of Education and Science, on 30th May, 1974. This points out that the HAC *"problem is not, as originally thought possible, confined to buildings with abnormally high roof temperatures, and humidity."* The circular further states: *"In addition, several other instances have been reported in which examination of buildings has revealed cause for concern about the safety of roof beams.*

It follows from this that there is now greater cause for concern, and that all buildings with roofs of a type similar to the Stepney school, that is to say having isolated prestressed beams of high alumina cement concrete, whether described as beams, joists or purlins, must be regarded as suspect. Lists of a number of such buildings, many of which are schools, have already been obtained from some of the roof beam manufacturers, and details have been passed on to the appropriate local authorities. These lists are by no means exhaustive, and it is therfore important that any further action a local authority can take to identify such buildings should be put in hand at once. The owners should be notified immediately of any buildings identified in this way, and the Department informed.

In view of the risk to life involved in the collapse of roof beams with little or no warning, you are advised that your own buildings should either be taken out of use or made the subject of temporary safeguarding measures until engineering appraisals have been made to establish whether the margins of safety of the design have been significantly eroded by deterioration of the high alumina cement concrete. It is suggested that the appraisals should aim at classifying the building as:

a) safe for a specified period
b) a risk needing further investigation, leading to a. or c.
c) such an immediate risk that either:
i. permanent strengthening measures should be taken to supplement the existing structure of the roof.
or ii the roof structure should be replaced as soon as possible."

And finally: "*The use of high alumina cement concrete in other elements of structure, such as floors, foundations, columns etc., and in roof beams of a different type or used in a different manner from those in the Stepney school, is not at present considered to involve the same order of risk. The Building Research Establishment is considering the implications of such use as a matter of high priority, and further advice will be issued as soon as possible.*"

July 1974

Following the publication of the report of the Building Research Establishment[35], the Department of the Environment, again acting also on behalf of the Department of Education and Science, issued another circular[92], dated 20th July, 1974. The circular applies to England; a similar circular[93] was issued by the Scottish Development Department, although the Department believed that HAC concrete had not been used to any significant extent in Scotland in isolated prestressed roof beams, joists or purlins. Since then HAC trouble has been discovered in the swimming pool roof at a school in Wick, in a school building in Thurso, and in two schools in Dundee (see Section 14.4).

The circular[92] of the Department of the Environment states: "*The Department, having considered the report, wishes to draw attention to the following main grounds for concern:*

a. The factors which are believed to have contributed to the loss of strength are complex, and the possibility exists that not all the factors have been identified.

b. While the vulnerability of high alumina cement concrete to attack by alkalis has long been known, it is now evident that when it has lost strength as a result of conversion, it is also vulnerable to sulphates, and possibly chlorides, in the presence of water, e.g. from leaks and from persistent condensation.

c. High alumina cement is vulnerable to loss of strength from changes in temperature and humidity which may occur at any time in the life of a building, as for instance if the heating or ventilating system is changed or differently operated; or if the use is changed in such a way as to increase the moisture and heat produced. Comparable effects may occur if structural alterations are made. Highly stressed prestressed flexural members with slim flanges and webs are sensitive to loss of concrete strength. The sensitivity is even more acute when the members are non-composite, particularly if they are isolated.

d. The high concrete strengths normally adopted in prestressed design may not be maintained at that level throughout the life of buildings with structural members of high alumina cement concrete, and there is no way of accurately predicting the future extent of deterioration, or when or over what period of time it is likely to occur.

The Department has also received a number of reports of structural inspections and appraisals carried out in accordance with the advice in my circular letters dated 28 February and 30 May 1974 which indicate loss of concrete strength in varying degrees and in many forms of construction.

Lessons: *The Department's conclusions are that:–*

(1) all existing buildings incorporating high alumina cement concrete must now be regarded as suspect, at least in the longer term;

(2) *a further programme of action is required starting immediately with those buildings with the greatest risk;*

(3) *high alumina cement concrete should not be used for structural work in building until further notice. An amendment to the building regulations will be proposed accordingly."*

Indeed, a programme of action was prepared[92]. This extends the earlier action on precast prestressed isolated roof beams in conditions of heat and humidity (circular of 28th February, 1974 [90]) and on such beams regardless of environmental conditions (circular of 30th May, 1974[91]) to *"all remaining precast prestressed non-composite roof of floor members (and also columns) of high alumina cement concrete, dealing in the first place with those buildings with roof or floor spans exceeding 5 metres. Within this class, buildings with structural members which are of sensitive cross section or which are isolated, those subject to the more extreme or variable environmental conditions, and those where failure would constitute the greatest risk to public safety, should be appraised first."*

The circular [92] also proposes that, when the results of work at the Building Research Establishment on HAC concrete under all conditions have been obtained, recommendations will be made about:

"(i) precast prestressed composite structural members
(ii) precast concrete structural members
(iii) in situ concrete construction."

An estimate of the work involved is difficult but it seems that the isolated roof beam construction has been used in about 400 buildings. A larger number of buildings falls into the general non-composite roof or floor category.

Some information is given in the circular [92] on the procedure to be followed. First, the use of HAC has to be established, i.e. the relevant buildings have to be identified. The sources of information are: the cement makers who know the precast concrete manufacturers to whom HAC was supplied, and the precast concrete manufacturers who know the contractors and sites to which they supplied the structural members. This information is passed on to the local authorities who inform the building owners, but it is of course realized that records of this sort are not necessarily accurate or complete. Moreover, some precast concrete manufacturers have gone out of business so that a part of the records is simply not available.

Structural designers are also a source of information about the use of HAC, but they do not always know whether or not HAC was used. In some cases, the original design envisaged the use of Portland cement but similar concrete units made with HAC have been substituted. The extent to which the designer was in control of such a change would depend on his original specification. For instance, in the case of the Bennett Building at the University of Leicester (see Section 6.2), it appears that ordinary Portland cement was specified but that HAC was used, although it is not known if this was authorised.

Yet another source of information on the presence of HAC concrete in a building is the owner or the local authority responsible for building control.

Following identification of a building with a potential 'HACproblem', the building should be appraised. The suggestions in the circular[92] are as follows:

"(i) Once a building has been identified as a possible risk the first step will be for the owner, with professional advice, to decide whether to close the building; or if not whether interim safeguarding measures are adequate or necessary.

(ii) The next step will be an engineering appraisal of the structure to establish whether the margins of safety in the design have been significantly eroded by deterioration of the high alumina cement concrete. It is suggested that the appraisals should aim at classifying the buildings as:

(a) safe for a specified period;

(b) a risk needing further investigation leading to (a) or (c)

(c) such an immediate risk that either:

(1) permanent strengthening measures should be taken to supplement the existing structure;

(2) the structure concerned should be replaced as soon as possible."

It is stated that "*a simple chemical examination is suggested as the initial test. Short of extensive drilling and testing of cores, which would be a destructive method for many prestressed concrete members, there is no reliable and accurate way of determining the strength of high alumina cement concrete. Nor can any future variation in strength be reliably predicted, nor even the lowest strength to which it may fall. Thus, test results are not in themselves decisive, they merely provide a contribution to the engineering appraisal, which will have to be made on the basis of professional judgement in the light of identified and potential risks.*"

And further:

"(iii) On the basis of the engineering appraisal the decision has to be taken about the future of the building. Since the available testing methods do not provide an automatic answer, there will be a wide area for judgement. In any case of the slightest doubt, the decision should go in favour of safeguarding action, whether replacement, permanent strengthening or periodic reappraisal; in the latter case with or without temporary precautionary measures.

(iv) Local authorities are reminded of the powers relating to dangerous buildings and structures under Section 58 of the Public Health Act 1936."

The latter section refers to the right of the authority to demand removal of a building deemed to be dangerous.

An appendix to the circular of 20th July, 1974[92a], makes suggestions for testing HAC concrete. The point is made that the extent of conversion cannot even be guessed from the surface appearance of the concrete so that the inner portion of a suspect member has to be investigated, but it is admitted that: *"There is no well-established test for this purpose . . . "*

As far as visual inspection is concerned, we are told[92] that *"it is unlikely that there will be any apparent change in external appearance of the concrete"*. Nevertheless, *"it is advisable to inspect for any signs of distress, such as excessive deflections, lateral bowing, and cracking (either shear cracking or flexural cracking). Special attention should be paid to any areas which have been subjected to moisture, due, for example, to leaks through roofs or to condensation. Sources of possible chemical attack, such as plaster and wood-wool slabs should also be identified. Both white and black marking may be indicative of chemical attack."*

The visual examination should be complemented by a chemical examination in order to determine the degree of conversion. Details of the procedure for taking of samples for this purpose are given in Section 12.2. The chemical test for the degree of conversion will also detect any chemical attack of HAC concrete in the areas sampled.

The real difficulty comes in interpreting the degree of conversion in terms of the strength of the concrete and, after all, it is the strength that we are interested in. The circular of the Department of the Environment[92] says: *"The degree of conversion increases with time in a complex manner but may be used as basis for indicating the quality of the concrete. On the basis of laboratory tests on high alumina cement concrete made with low water-cement ratios and kept at ordinary temperatures it has been observed that the minimum strength after several years does not fall below the one-day strength. The degree of conversion for such concretes is shown against time in Fig. 1* (Fig. 9.1 in this book) *curve A, and any concrete with an equivalent or lower degree of conversion may be considered as of 'good' quality. Information on degrees of conversion of the high alumina cement in some structures which have suffered collapse, chemical attack or marked reduction of strength have enabled a further curve to be proposed, i.e. curve B, parallel*

Figure 9.1 **Assessment of the quality of HAC concrete on the basis of its age and degree of conversion, according to the Department of the Environment[92].**

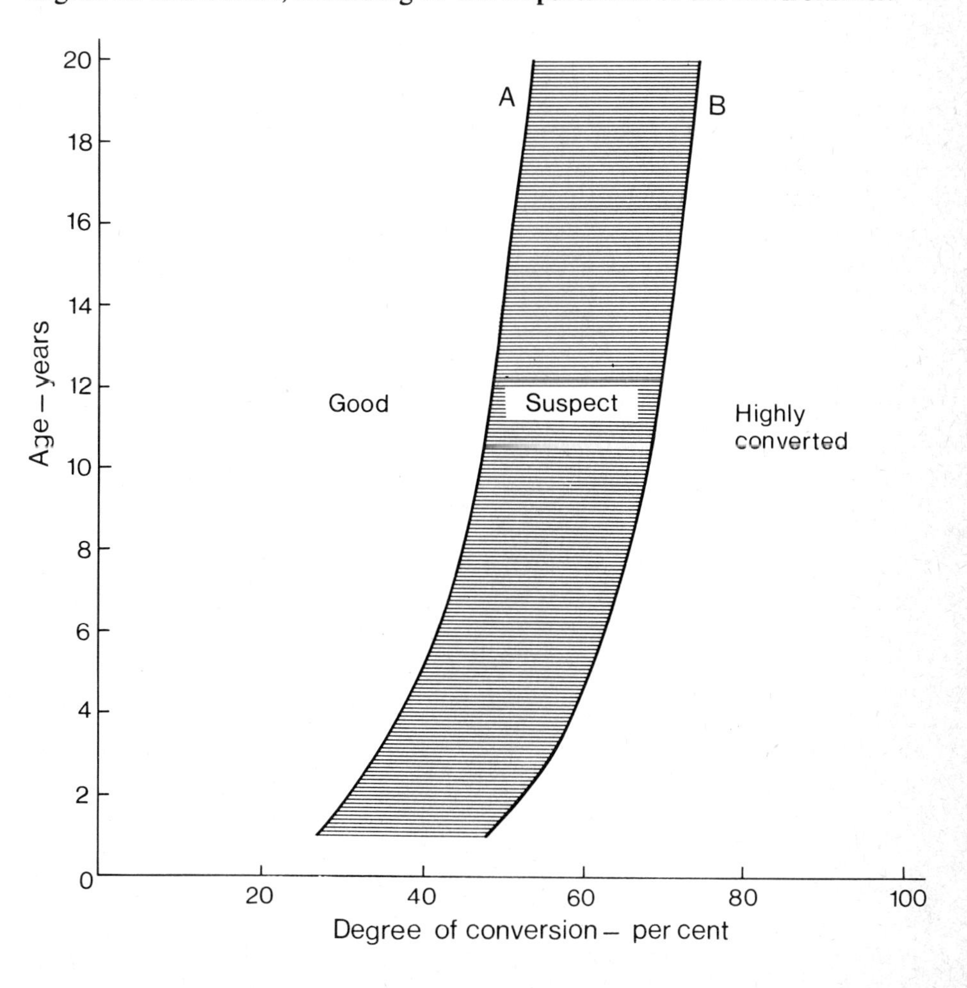

to curve A. It is suggested at present that all results with degrees of conversion greater than those shown by curve B should be interpreted as indicating 'highly converted' concrete. Results falling between the two curves would indicate some excess conversion, and possibly some loss in strength, and should be regarded as 'suspect'.

'Good' quality concretes are likely to give satisfactory service, but it would be prudent to re-inspect at intervals. If there were any change in the environmental conditions around the concrete, it would be necessary to re-test at intervals.

'Suspect' quality concretes are likely to give satisfactory service in the short-term, but may not in the long-term, so that it would be important to test again at intervals. Tests should be carried out at shorter intervals for beams up to 10 years of age. Further types of tests may help to identify more closely the properties of the concrete.

'Highly converted' quality concretes are much less likely to be satisfactory in service. Such concretes are vulnerable to chemical attack, unless no materials containing aggressive chemicals are in contact with the concrete. Chemical attack, where it occurs, is likely to be a long-term activity and where members are retained as part of a structure inspection would be advisable at intervals. Furthermore, it is likely that the minimum strength of such concrete will be equivalent to the strength of 'fully-converted concrete' in relevant Codes of Practice. There is now evidence that strengths may fall below these values, in particular where high water-cement ratios occur. Again, further types of tests may help to identify more closely the properties of the concrete."

Two points in the last paragraph should be emphasized. One is that highly converted concrete is not likely to be satisfactory in service. This may be a somewhat pessimistic view in somes cases but, in others, fully converted concrete has been seen to be soft, crumbly or mushy. The beams at the Cadbury factory fall into that category (see Section 14.3), and so do some of the beams at the University of Leicester, examined by the Author. The second point is that the actual strength may be lower than would be expected from the data in the codes of practice. As we have already said (see Chapter 7), it is the expectation that was wrong, and not the concrete that behaved abnormally. In other words, in many cases, we should expect conversion to lead to extremely low strengths.

Core test

Further tests, involving a progressively increasing effort and of greater complexity in their interpretation, are reviewed in the circular[92]. The first of these is the core test. Details of core testing are given in Section 12.4 but it should be emphasized here that core drilling is a skilled operation and that the interpretation of the test result is by no means simple or unambiguous.

Ultrasonic pulse velocity test

The second test discussed is the ultrasonic pulse velocity determination. The most important feature of this test is that it is solely of a comparative nature. The test does not yield a direct indication of the strength of the concrete because the relation between the pulse velocity and the strength depends on the type of aggregate used and on its quantity, as well as on secondary factors such as the moisture content of the concrete and the presence and direction of reinforcing (or prestressing) steel. The subject is discussed in Section 12.3.

Loading test

Finally, the circular[92] mentions the possibility of a load test. This would follow the usual engineering practice (see Section 12.5) but it is suggested that the accepted value of load equal to 1.25 times the live load *"may not be adequate for a material which can lose strength with time. A load of 1.25 times live load plus 0.25 times dead load is considered the minimum appropriate."*

All the tests mentioned are directed, implicitly or explicitly, towards an estimate of the compressive strength of the suspect concrete. We assume that, as in the case of Portland cement concrete and of unconverted HAC concrete, there exist fixed relations between the compressive strength on the one hand and the tensile strength and bond strength on the other. Indeed, we assume that these relations are not affected by conversion so that from an estimate of the compressive strength of converted HAC concrete we can deduce the likely tensile and bond strengths. Some confirmation of this is given by the tests reviewed in Section 4.1, and the assumption of the usual relation between the compressive and other strengths is believed to be sound (see Fig. 4.4).

One further point made in the circular[92] of 20th July, 1974 should be mentioned. This is a statement that, for some members, a reduction in the strength of concrete could lead to a reduced period of fire resistance. We can only guess that this is related to the strength of the converted concrete and its porosity and hope that, in view of the importance of the fire resistance, a further elucidation will be forthcoming.

Following the amendment[87] to the Building Regulations abolishing the deemed-to-satisfy status of HAC, the Department of the Environment issued a further circular[128] on 2nd January, 1975. This explains that local autorities will be able to reject proposals involving the structural use of HAC, and states:

"There could, however, be circumstances in which local authorities would wish to consider allowing the use of the material to continue, and the following recommendations are made accordingly:

a. Structural members
The Department considers it inappropriate in general to approve proposals involving the use of structural members made with high alumina cement. Where resistance to sulphate attack is required, it is preferable to use sulphate resisting Portland cement.

b. Lintels
Where loads from floor joists, trusses, beams and rafters are to be supported, lintels of other material should be used. It is not however at present considered necessary to suggest a total prohibition on the use of concrete lintels made with high alumina cement with spans of less than 2 metres where composite action with brickwork or blockwork will develop.

c. Flue blocks and refractory uses
The Department does not consider it necessary to restrict the use of high alumina cement in flue blocks or for refractory purposes, but account should be taken in design of losses of strength which will occur in service.

d. Foundations
It would be prudent not to use high alumina cement concrete in foundations where alternative materials are practicable."

The same circular[128] states that in March 1975 the Department of the Environment will give advice *"on whether or to what extent appraisal and testing should be extended"* to *"precast prestressed non-composite structural members used in buildings with roof or floor spans not exceeding 5 m"*.

Later in the year, advice will be given on *"precast prestressed composite structural members, precast concrete structural members, and in situ construction"*, where *"the risk of sudden collapse appears to be small"*.

PSA Instruction

Although this is not a document for public use, a technical instruction[94] issued by the Property Services Agency of the Department of the Environment is of interest as it reflects the attitude of the government agency responsible for a large number of buildings. It may be worth noting that even before the present 'HAC troubles' erupted, the Property Services Agency allowed the use of HAC only with an explicit permission of the Superintending Officer responsible for the building. In fact, HAC was not used in any of the buildings designed for the Agency, except when they were specified by what is known as a performance specification or where the 'design and construct' approach was used.

The instruction[94] was issued in August 1974 and it simply bans HAC from use in the construction of the Property Services Agency buildings except for non-structural linings or where sulphate resistance is required in foundations. But even in those cases prior approval of the Directorate of the Civil Engineering Department is necessary.

Much of the instruction[94] echoes the circulars of the Department of the Environment referred to earlier, but one or two points are of interest. It is said clearly that: *"It is not prudent to accept without question the manufacturer's design water-cement ratio as in practice the water-cement ratio achieved, which should not be greater than 0.4 may be higher either generally in the member or locally."*

This is a very sensible recognition of the fact that the manufacturer's maximum water-cement ratio cannot in reality be a guaranteed value.

Another statement tells us clearly that *"conversion of HAC leaves the concrete in a porous condition and vulnerable to chemical attack"*.

We should also note the definition of composite construction, which is less vulnerable than isolated beams: *"Composite construction as used in this Technical Instruction means that the structural members have been designed to act in conjunction with a reinforced structural topping with shear links. Note particularly that a mesh reinforced sand/cement screed should not be regarded as composite construction. In this context also HAC concrete joist and hollow pot floors or roofs are regarded as non-composite construction but in a lower risk category to isolated non-composite beams."*

This is a realistic appraisal of the situation, and at variance with some claims which we encountered when inspecting several buildings: there, hollow pot floors and a screed were regarded as acting compositely. We shall look at the matter again in Chapter 14.

10. Manufacturers' Attitude to High-Alumina Cement Conversion Problems

The various long-term tests discussed in the preceding sections add up to a clear picture of the long-term behaviour of HAC concrete. Were the cement makers and manufacturers of precast concrete aware of the situation?

The answer seems to be yes. In a paper[6] written by the technical director of Lafarge Aluminous Cement Co. Ltd. and published in March 1959 in Australia it is stated that *"a reduction in strength can take place, but at a much slower rate (than when 'overheated' during casting) if normally matured HAC concrete is, in subsequent use, maintained for long periods in water above 80°F to 85°F (27°C to 29°C). For this reason it is usually agreed that high load-bearing capacity should not be demanded of HAC concrete, for example, in tropical seas where these storage conditions might be realised. It would appear that both saturated conditions and high temperatures are necessary together for the conversion to take place, since there are many old structures above water which are in excellent condition after being subjected to climates noted for high temperatures."* This is possibly an over-optimistic conclusion.

An investigation[34] carried out at Imperial College at the request of the cement manufacturers gives quantitative data on conversion but these can be seen to be unduly optimistic. Figures 10.1 and 10.2 give the converted strengths for various water-cement ratios, and these can be compared with the data of Figures 3.14 and 3.17.

Some of the precast concrete manufacturers went even further in minimizing the importance of conversion, claiming to do so on the basis of the cement makers' research. This is evidenced by the following statements in a paper[36] published in April 1961: *"Appreciable conversion, in the climatic conditions of Great Britain, seems improbable even over a long span of years for properly cured concretes."* And further, *"strength deterioration in moist conditions is negligible if temperatures are kept normal, i.e. 64°F (18°C)."* The paper concludes *"that HAC concrete may safely be employed in this (British) climate provided that maintained hot and wet conditions such as in laundries, boiler houses and cooling towers are avoided and provided that proper precautions are taken when casting."* The implication that if the hot and wet conditions are not maintained there is no trouble is in marked contrast with our 1958 paper[33] in the Journal of the Institution of Civil Engineers, and the whole tenor of the precast conc ete manufacturer's paper is that trouble can be expected only in rare applications. This attitude is unrealistic. The paper [36] reports that *"several million square yards of such*

Figure 10.1 **Strength of HAC concrete of various mix proportions stored in water at 38°C from the time of casting (102mm cubes)[34].**

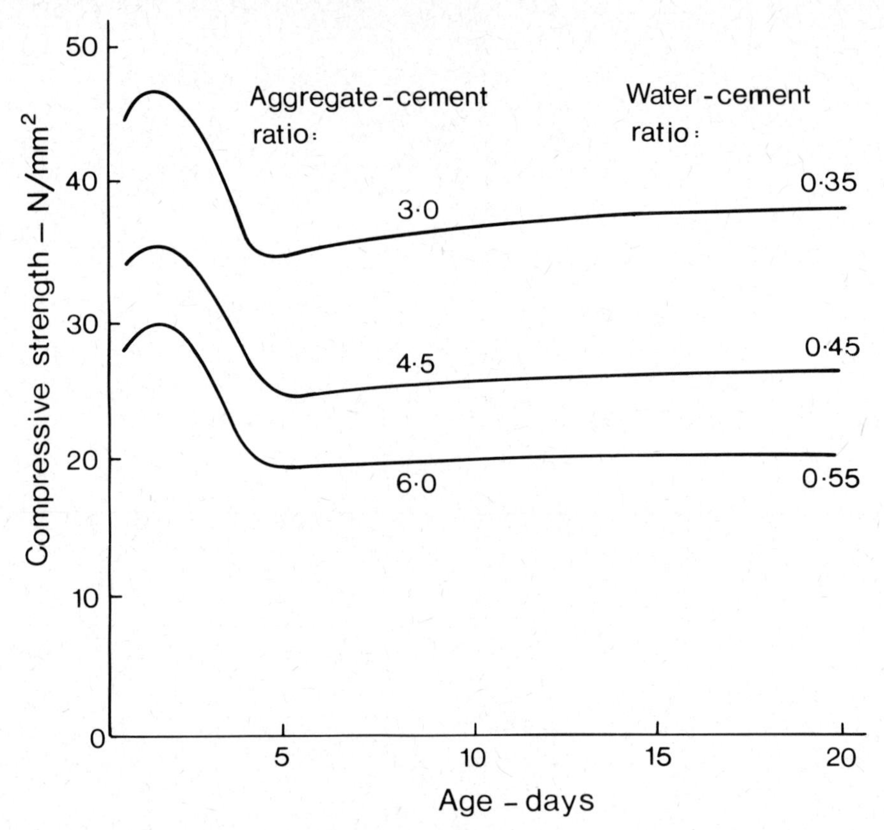

flooring have been put into use during the past ten years in this country and no trouble has been reported" and it is claimed that *"care is taken that such units are not used in certain industrial situations or places where hot, wet conditions will be maintained."*

Another quotation from the paper[36] is of interest: *"It seems therefore that there is no reason for anxiety about HAC structural products made, and used inside buildings, in this country provided proper precautions are taken to prevent excessive temperature rise during the curing period. This is an easy matter in our climate. . . "* And further, *"Manufacturers of precast HAC products have no difficulty in keeping curing temperatures down . . . "* May we add that the manufacturer for whom the author of the paper had been the chief engineer made the beams for the Stepney school.

There is a further surprising statement in the paper[36]: *"Lafarge also says that conversion requires time, as well as heat and moisture, and that internal temperatures up to a maximum of about 130°F (54°C) can be accepted provided the excess temperature is dissipated within 24 hours or so; in such cases the concrete can be regarded as normal."*

Another mention of the work of Lafarge in the same paper[36] speaks of *"the unlikelihood of conversion occurring in properly matured concrete even during the occasional warm spells that occur in this country."* Lafarge is reported as having made cubes, normally cured and put out in the open for five years. *"Examined in January 1961 by both X-ray and chemical techniques these showed no trace of conversion."* [36]

Two years later, in the discussion[126] of our 1963 paper[45], the same manufacturer (Mr. O. J. Masterman) wrote:

Figure 10.2 **Strength of HAC concrete of various mix proportions cured in air at room temperature for 6 hours and thereafter stored in water at 38°C (102mm**

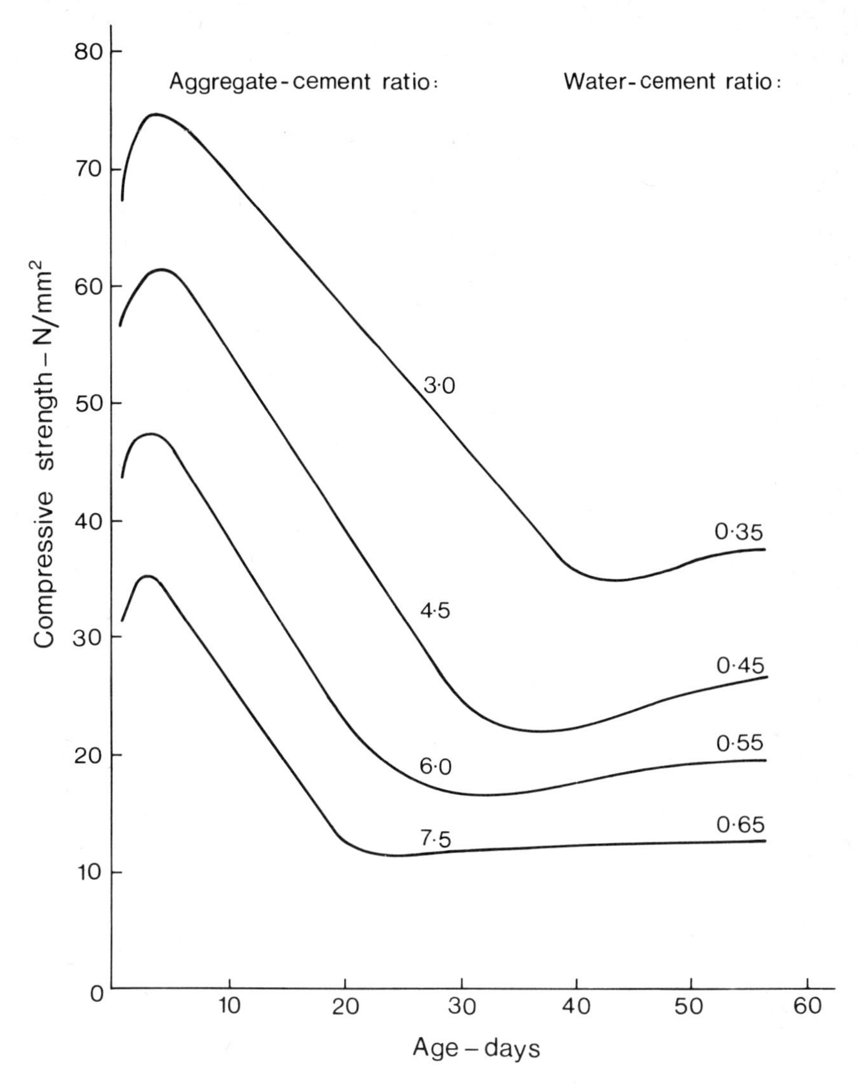

"The Author had concerned himself too much with reported failures and too little with the successful use of high-alumina cement. It would be easy to compile a much larger list of failures in concrete structures made with Portland cement than he had compiled for high-alumina cement, but Mr. Masterman felt that the Author had not had enough contact with the consultants and firms who were using this material correctly, and who knew what precautions to take." He further *"took exception . . . to the (Author's) statement . . . that at ordinary temperatures prevailing in England conversion took place spontaneously". He went on to say: "On the second point, that high-alumina cement concrete spontaneously converted at normal temperatures because it was unstable, this might equally be said of another common building material, glass, which also was unstable in the chemical sense. It was a question of how soon. One did not worry about the glass, and in Gt. Britain one need not worry about high-alumina cement concrete which was properly made and used in suitable conditions."*

In the same discussion[127], the managing director of Lafarge Aluminous Cement Co. Ltd. (Mr. J. T. Kay) said:

"The Lafarge Aluminous Cement Company took the strongest exception to

imputations that wrong information on cement properties had been supplied knowingly or that information had been withheld. It was clearly against the interest of the manufacturers to have their material used improperly or in unsuitable conditions. The Company had long maintained a highly competent Technical Service Section, which had been widely used; at no time had established information been withheld from any interested person or organization."

Early promotional literature by the cement makers makes no reference to conversion or loss of strength but in their 1961 brochure[95] they explain the *"reason for preventing overheating "*. Conversion is mentioned, albeit in quotation marks, and we are clearly told that HAC *"is not recommended for high load-bearing* structural concrete where conditions of both high humidity and high temperature occur simultaneously (e.g. in tropical climates or certain industrial applications) although in hot* "dry" *climates (e.g. Algeria)* (HAC) *concrete has been shown to be in excellent condition after 25 years or more. In temperate climates* (HAC) *is, of course, used quite satisfactorily."* Presumably the latter statement refers to the United Kingdom. The brochure reproduces the graph from Ref. 34, and points out the dependence of conversion effects on the water-cement ratio.

The effect of conversion on porosity is clearly mentioned, and it is recommended that for chemical resistance work at normal temperatures the one-day *"preliminary cube strength"* should be not less than 45 N/mm^2. Overheated concrete is said to be not necessarily of sufficiently low porosity. Where conversion must inevitably occur we are told to use a water-cement ratio of 0.35 to 0.40 and a *"somewhat richer mix than normal"*. This advice is said to apply in the case of linings of tanks containing hot liquors, gas-washing plants, precipitators, chimneys etc. However, there is no hint that conversion may and indeed must occur under much less onerous conditions.

The advice on piles given in the brochure[95] is not really sound. We are told that the usual mix for concrete piles or extensions is 1:2:4 and *"leaner mixes are not generally recommended where chemical resistance is also required unless vibratory compaction is employed so that full consolidation can be obtained with reasonably low water-cement ratios (e.g. 0.5)"*.

On the other hand, the recommendations for pretensioned concrete units are unexceptionable: *"a water-cement ratio of no more than 0.35 to 0.40"*. This is in contrast to normal use where *"a 1:2:4* (HAC) *concrete mix to be placed by hand normally requires a water-cement ratio of 0.5 to 0.6. When vibratory compaction is employed this may be reduced to 0.4 and 0.5."* These values refer to the total water and not to the effective water; on the latter basis, the values of the water-cement ratio would be 0.05 to 0.10 less than quoted.

Some misleading statements, published as late as 1962[3], may be worth quoting: *"An orthodox HAC concrete (1:6), with a consistency suitable for consolidation by hand, will usually require total water-cement ratios of 0.55–0.60. If well-cured it will readily give high strength, high chemical resistance, low porosity and low permeability. It is therefore eminently suitable for structural work as long as it is not submitted to conditions causing relatively quick conversion."* The reader who wonders when conversion will take place is told that *"Even when the concrete is maintained continuously under water, conversion is not of practical consequence when*

*Was high-load bearing concrete really meant?

the water temperature is below about 25°C." One could point at the Stepney school roof which was above water until it could no longer support itself! We are further assured that *"Long-term tests on large numbers of HAC concrete specimens exposed to the elements in temperate climates (e.g. in Great Britain) show that any subsequent conversion can be of little practical consequence in the case of well-made and properly cured concrete."* This statement is supported by a reference to Lafarge Aluminous Cement Co. Ltd. Laboratories, and appears in a book by its technical director[3] .

Later information put out by the cement makers is much more conservative. In July 1973[96], we are told that prestressed concrete should have a *"free water-cement ratio less than 0.40 and normally less than 0.35".* Concrete with a free water-cement ratio of less than 0.35 is said to *"retain good strengths (over 25 N/mm^2) even when fully converted".* This is undoubtedly true but is it possible to ensure, even in a factory, that the outer limit of the water-cement ratio in every unit made and in every part of every unit be 0.35? We believe that the answer is no, i.e. that inevitable variation will lead to a higher water-cement ratio here and there, and the consequences of this are much more serious than when Portland cement is used. That water is not always batched accurately is illustrated by a promotional film made for the cement makers which shows, at a precast concrete factory, water being manually poured into the mixer by means of a jug filled from a 45 gallon drum.

Although HAC is made in the U.K. by one manufacturer only, it was marketed, in addition to direct sales, by the Blue Circle Group. Their advice was cautious. In a technical note published in December 1971 it is said that *"Provided that the amount of water used to hydrate the cement was sufficient but low in total, adequate strength and chemical resistance will remain after conversion."* This is followed by an exhortation: *"Detailed study of the following paragraphs and efficient production control are, however, necessary to ensure that control of water-cement is at the correct level."* We are also told that *"Conversion can occur at any time during the life of the concrete"*, and *"It is necessary . . . to allow for the full loss of strength . . . in all cases where the consequence of not doing so might be serious."* This seems sensible.

Let us revert to the cement makers. In 1968, the technical manager of Lafarge Aluminous Cement Co. Ltd.[97] wrote: *"At the temperatures normally met in this country the use of high-alumina cements rarely poses any great problems."* This is rather an understatement. He continues[97] : *"The effects of the hydraulic activity and heat release in the first 24 hours are as follows.*

(1) Moisture tends to be driven towards any exposed surface, and if its loss is not prevented the surface may dry, fail to harden properly and be weak and dusty.

(2) Surfaces that dry too quickly may shrink abnormally, while underlying concrete that retains its moisture also retains its normal shrinkage characteristics. Such a difference in shrinkage sometimes causes flooring to curl or crack, particularly if the bays are too thin, too large in area or badly bonded to the base.

(3) If the mass of concrete is large enough, and the heat is not dissipated by cooling, the temperature within the hardening concrete might mean above 27°C (80°F) for more than 24 hours. Nevertheless, provided it had been well

compacted at a total water-cement ratio below 0.5 the strength and impermeability remaining in the concrete would suffice for many purposes. Internal temperatures of above 38°C (100°F) are virtually harmless, provided they are reduced below the above limit within 24 hours of placing." The expression "*virtually harmless*" may be noted.

And further: *"As long as the work is no more than a few inches thick (as would usually be the case in road or floor construction), it will cool to atmospheric temperature well within the first 24 hours, and (3) above may for practical purposes be ignored. In these circumstances the prime purpose of water-curing is to retain moisture, and techniques common to ordinary concreting (e.g. applying membrane curing compounds or covering with impervious sheet such as polythene) may be used."*

Concreting in hot weather with HAC is suggested to be not much of a problem. The paper[97] continues *"To produce normally hardened high-alumina cement concrete in really hot weather can sometimes demand measures such as shielding aggregates from the sun, hosing down plant and shuttering to cool them, reducing the depth of the lifts and not only applying cooling sprays to the concrete but also shielding it from the direct sun during and after hardening."*

We should note that measures are demanded only *"sometimes"*.

Hot countries are thought to present no serious difficulties[97]. *"In countries where the water temperature may remain above 80°F (27°C) for long periods, it may be impossible to produce normally hardened high-alumina-cement concrete, yet even in these conditions the material is successfully used by adopting a very simple technique.*

In brief, this technique is based on the relatively recent discovery that 'overheated' (or 'converted') high-alumina-cement concretes and mortars can have excellent strengths and high impermeabilities. This is provided that they are designed for full compaction at total water-cement ratios of less than 0.4 and then kept moist (not cooled) during hardening.

Because of the high workabilities of high-alumina-cement mixes at low water-cement ratios, it is possible to do this by relatively small increases in the normal cement contents. The cost of these is often less than that of the precautions needed to ensure normal hardening in hot climates.

An added advantage is that the concrete produced merely improves with exposure to prolonged simultaneously hot and humid conditions. As is well known, these used to be considered highly unfavourable to all high-alumina cement concretes, irrespective of their water-cement ratios."

Even the July 1973 issue of the cement manufacturers' data sheet[96] on mix proportioning (S2) says: *"If concrete is cured with water the concrete is cooled and the concrete will not convert unless the section is very thick."* But there is a proviso that *"no liability or responsibility of any kind (including liability for negligence) is accepted in this respect by the company, its servants or agents"*.

11. The Press and the High-Alumina Cement Problem

Birmingham Evening Mail

Because the HAC troubles have erupted from their engineering confines, some of the newspaper comments are of interest. On 30th August, 1974, the *Birmingham Evening Mail* wrote under a title *Danger cement warning ignored: "Birmingham's Conservative education group has asked the Prime Minister for an investigation into why a warning against using quick-dry cement in the construction of schools and other buildings was ignored.*

This follows the disclosure that top experts warned against the use of 'high alumina' cement over 12 years ago, and that its use has been banned on the Continent.

At the moment, 60 buildings, including nine schools, have been the subject of top priority safety checks by Birmingham engineers and architects."

And further, it is stated that Councillor Neil Scrimshaw *"revealed that a top expert in building construction, Professor Adam Neville, head of the civil engineering department at Leeds University, had previously warned against its use.*

Sunday Times

A spokesman for the Department of the Environment admitted that they were aware of the warnings made by the professor several years ago.

It was also known that its use had been banned in other countries. 'But other counsels prevailed', said the spokesman."

If this statement is correct, then it would be interesting to know what these counsels were. Anyway, they were there, and, even now, after the report of the Building Research Establishment[35], still are there, for, according to an article by Anthony Holden in the *Sunday Times* of 21st July, 1974, "Mr. *Leon Grice, marketing director of Lafarge, said yesterday that . . . (the company) stood by its claim that HAC was perfectly safe if the British Standards Institution code of practice was strictly observed."*

The Surveyor

The re-discovery of our old warning is mentioned also in *The Surveyor* of 19th July, 1974 which published an editorial staff report entitled *High-alumina cement—1963 warning ignored?* This starts by saying: *"While reports continue to roll in from different parts of the country of school buildings being closed until the safety of their roof-beams can be checked, we have been reading a paper by Professor A. M. Neville (then of the University of Alberta but now Head of the Department of Civil Engineering, University of Leeds) published in July 1963 in volume 25 of the Proceedings of the Institution of Civil Engineers.*

Professor Neville's paper 'A study of deterioration of structural concrete made with high-alumina cement' collates the results of laboratory investigations and seeks to establish that there was a gradual loss of strength in the high-alumina cement concrete investigated. It goes on to consider the factors behind this loss together with the structural changes on conversion of hexagonal hydrated calcium aluminates to a cubic form. The paper also reviews structures made with this material in Britain and overseas and reports restrictive regulations introduced in a number of countries.

Described in the written discussion by C. D. Crosthwaite (partner Freeman, Fox & Partners) as 'one of the most important that the institution had received for some time', Neville's paper was heavily attacked by trade figures including O. J. Masterman (Director, Pierhead Limited), author of two publications referred to in the Building Research Establishment current paper on the Stepney School collapse, published in June this year.

Though there is no real substitute for reading both the paper and the written discussion (published in May 1964 in volume 28 of the ICE Proceedings), we are grateful to the ICE for permission to publish some extracts which we hope give the general gist."

After a full description of the contents of the paper, the article reviews the discussion. This is quite revealing and bears quoting: *"Though the author of the paper did not appear to be rattled by the discussion, it does represent a very powerful and sometimes persuasive attack on the basis of his work. O. J. Masterman of Pierhead Ltd. felt that in reporting loss of strength in concrete, 15, 20 and more years old, he (Neville) was resting his argument on concretes made at a time when this special type of cement was certainly not well understood, when precautions which would be taken today were not practised, when water contents were unquestionably too high and when mechanical vibration was not usual.' The spontaneous conversion of high-alumina cement concrete at normal temperatures need not worry Gt. Britain.*

J. T. Kay (Managing Director, Lafarge Aluminous Cement Co. Ltd.) believed that the author 'had little or no experience in the practical use of high-alumina cement but, unfortunately, this has not prevented him from assuming, at times, the position of arbiter on such matters.'

Jan Bobrowski (Consulting Engineer) reckoned that the author's distinction between Portland cement and high-alumina cement was very useful. On the other hand, he found the paper very misleading: 'first it implied that a full range of high-alumina cement concretes, which could be used structurally, was covered by experiments and tests reviewed in the paper, and secondly, by systematic tabulation of structural failures in Great Britain and abroad it created the impression that at least most of these could be attributed to conversion of high-alumina cement. Actually the ignorance of many users of this cement, and even negligence, would often be the cause.'

Bobrowski said that the water-high alumina cement ratio should be as low as possible and certainly not more than 0.4. This made efficient mechanical vibration necessary with subsequent testing not only of strength but also of density to ensure compaction. He recommended a very low cement content: 'most suitable mixes would not be much richer or leaner than 1:8'. In conclusion Bobrowski assured the author that with the mix he had described and a requirement that the 24 hour crushing strength should not be less than 9000lb/in^2 (62 N/mm^2), 'a high-alumina cement concrete, under normal *conditions in Great Britain, would gain rather than lose strength. It would also keep gaining it even if artificially converted at an early stage.' He had seen plenty of evidence to this effect and the confirmation could readily be obtained from reputable manufacturers in Great Britain. He was also confident that, with the mix given above, used properly, the ultimate design could be based on 9000lb/in^2 (62 N/mm^2) strength with complete safety.*

J. K. Sykes (Sir Robert McAlpine and Sons Ltd.) wrote that in his experience conversion rarely occurred in Great Britain. 'Furthermore, in the instances when loss of strength is encountered, such loss is often attributable to other causes.' To his knowledge, 'high-alumina cement has been used entirely successfully and extensively over a period of many years in prestressed concrete beams '.

T. D. Robson (Technical Director, Lafarge Aluminous Cement Co. Ltd.) believed that hardening at high temperatures would show lower strengths. This conclusion had not been queried since shortly after Freyssinet and Coyne had advanced it in 1927 . . . 'there was no doubt that submission to hot wet conditions (certain industrial applications, tropical waters etc.) would cause a relatively rapid rate of conversion and a fully converted concrete could then be expected within a practical period of time. A high-alumina cement concrete likely to be satisfactory in such conditions could be obtained only by the use of low water-cement ratios, and therefore high-alumina cement concrete should not be considered for load-bearing structural duties in these conditions without prior reference to the manufacturers '.

E. A. Tetlow (Director, Trent and Hoveringham Concrete Companies) submitted that tests by his company had shown that 'with water content not exceeding 0.38, even if conversion took place, the deterioration of the concrete in terms of strength was normally within a tolerable amount '.'So far as his experience went (and it involved the manufacture of possibly half a million pretensioned beams), there could be no doubt that if the concrete was made in accordance with the recommendations of the new draft code of practice for pre-cast concrete, conversion did not take place during manufacture.' Either water or heat or both appeared to have been a factor in all the cases reported in the paper and it would seem that the use of high-alumina cement must be avoided where conditions of heat and humidity were likely to arise. In the vast majority of cases involved in the use of pretensioned beams 'the concrete was not exposed to damp and in fact was insulated in most cases from heat by screed or plaster finishes '.

Apart from contributions by T. N. W. Akroyd and C. D. Crosthwaite, both consulting engineers, and two speakers from the Department of Mines & Technical Surveys, Ottawa, Canada, all of whom gave the author some support, the remainder of the discussion was taken up by two contributors from Ciments Lafarge Co., Paris. They wrote that Ciments Lafarge regarded conversion 'as a known reaction, the results of which need have no practical consequences'. Observations were produced demonstrating that the strength of concretes containing the hydrate in cubic form 'continued to increase with time '.

Doctor G. H. Sadran (Director of the Central Laboratories, Ciments Lafarge) accepted that there was little disagreement over the conversion of hydrate. Their different interpretation rose from the fact that the author limited himself mainly to water-cement ratios at levels 'which rightly were no longer valid for structural high-alumina cement concretes while Ciments

Lafarge were dealing with the lower values which were recommended. In the latter case conversion lost a serious aspect which was formerly attributed to it and was regarded as a known reaction, normal in certain circumstances, and without practical effect '.

In reply the author gave as good as he got and it appears there the matter was left—until the roof beams failed at the Sir John Cass's Foundation and Redcoat Church of England School in Stepney, London. Now as if with some blinding revelation all Britain is looking at its pre-stressed roof beams made of high-alumina cement concrete. We find it very odd that so little notice has been taken of Professor Neville's paper and its discussion in the decade following the publication in the Proceedings of the Institution of Civil Engineers."

Building

Building of 19th July, 1974 carried an article called *High alumina: the facts were known ten years ago.* This starts by saying: *"Strong warnings on the loss of strength that could emanate from using high alumina cement concrete structures were given more than ten years ago by Professor Adam Neville, who was recently appointed President of the Concrete Society. Moreover, consultant engineers should have been well aware of these views since they were contained in a paper published by the Institution of Civil Engineers. This was in July 1963."*

A further note in the same paper gives such a boost to the Author's ego that it cannot be quoted.

The Times

In *The Times* of 7th October, 1974, Malcolm Brown put the issue of past warnings rather well: *"Abstracts from the proceedings of the Institution of Civil Engineers are not what the layman would normally consider compulsive reading but, with the benefit of hindsight, two sentences on page 84 of the May 1964 proceedings stand out with unquestionable clarity. They are taken from a letter written the previous year by Professor H. Rüsch of the Technical University of Munich to Professor Adam Neville of the University of Alberta, who was a leading figure in the fight to have it recognised that high-alumina cement had inherent dangers.*

Choosing his words carefully Professor Rüsch wrote: 'As I learnt, in England the conclusion has been drawn that high-alumina cement concrete can be used in prestressed concrete when the precautions . . . are observed; one will only allow for the possible loss of strength in the design calculations.

I fear that in England, too, this will lead to serious consequences.'

Just what Professor Rüsch's 'consequences' were is now being realized—with a vengeance. Around the country thousands of buildings are having to be examined to ascertain whether high alumina cement (HAC) was used in their construction. The material has effectively been banned by the Department of the Environment."

The article ends by saying that the *"use* (of HAC) *has always been controversial.*

It seems fair to ask, in the circumstances, where we lost our perspective on the matter. There was plenty of evidence suggesting caution. The warnings were sounded clearly enough. Our failure to hear them may have been an extremely expensive mistake." This problem is considered in Chapter 14.

Yorkshire Post

A similar theme is followed in the *Yorkshire Post* of 22nd July, 1974 under the title *Professor drew attention to cement risk—11 years ago: "The paper on the cement by Prof. A. M. Neville, head of the Department of Engineering at Leeds University, was published in July, 1963, in the Journal of the Institution of Civil Engineers.*

Described at the time 'as one of the most important received by the institution', the paper collated results of laboratory investigations on high-alumina cement and concluded that 'the repeated evidence of the gradual decrease in strength over long periods is irrefutable'.

Mr. Thomas Akroyd, a London consulting engineer, who supported Prof. Neville's comment in the discussion which followed his paper, said yesterday that there had been warnings about the problem of high-alumina since 1956, including warnings by Government establishments.

He thought the Government was acting responsibly by ordering checks on buildings and added: 'What we should be asking is should designers have used high-alumina knowing what they did about it'. The information has been available for a long time."

And, despite all this, the article reports that *"The marketing director* (of Lafarge), *Mr. Leon Grice, has stated that the company stood by its claim that the product was perfectly safe if the British Standards Institute codes of practice were strictly observed."*

New Civil Engineer

There were a great many other newspaper articles and they all express surprise at the 'HAC problem' in our structures. On the other hand, the *New Civil Engineer* has remained serene about HAC throughout. Conversion is probably not to blame, it feels. On 14th February, 1974, less than a week after the Stepney school collapse, it wrote that the *"determination of the precise mechanism of failure has all the makings of a major chemical mystery"* and *"Bobrowski remembers boiling specimens in water in that period and says of the design that 'it was made in the full knowledge of what was known about conversion of high alumina cement'. 'Even if there had been 100 per cent conversion the beam would still be adequate.'"*

The Surveyor

In the 22nd February, 1974 issue of *The Surveyor* Jan Brobowski is reported as saying, *"Without the slightest reservation I can state that there is nothing structurally wrong. I am convinced that alkaline hydrolysis of the cement is to blame."*

The same article in *The Surveyor* also reported that, *"Lafarge Alumina Cement, the French company which provides the materials for the Pier-head beams, was equally certain that conversion, the crystalline structure change which was one of the factors in the Camden collapse, was not involved in this case. Chemical attack of some sort was the cause, and the company's literature pointed out that high alumina cement was subject to attack by caustic alkalis, said L. G. Grice, marketing director."*

Even after an interim report on the Stepney school had been released by the Building Research Establishment, we were told (in *The Surveyor* of 14th June, 1974) that *"Jan Bobrowski, consulting engineer and designer both of the pool and the roof beams, said this week that it was important to bear in mind that this was only an interim report. 'I still say there was nothing structurally wrong.'*

He said he over-designed the beams to withstand a higher stress than the Code of Practice laid down. 'Unless the Ministry in its final report can actually prove that a properly made high alumina cement concrete, with all the safeguards being taken,can deteriorate to these scandalously low levels, I can only think that the possible explanation to the collapse must be one of the three factors still being investigated.' "

New Civil Engineer

On 21st February, 1974, the *New Civil Engineer* made a remarkable statement that *"now that the process* (of conversion) *is largely understood, the material* (HAC) *may safely be used under almost any environmental conditions."* And on 7th March, 1974, under the title *HAC may be innocent scapegoat,* John Parkinson wrote in the *New Civil Engineer: "The most surprising aspect of the latest DoE warning and appraisal advice to local authorities is that, in the absence of a specific explanation for the Stepney collapse, the BRE* (Building Research Establishment) *is directing investigators to search for conversion and playing down the possibility of chemical attack.*

This can be explained by the attitudes previously expressed by the BRE towards high alumina cement. It is also consistent with the BRE line of logic used at Camden, where several engineers feel that this cement was used as something of a scapegoat."

The article further says that at water-cement ratios below 0.4 the effect of conversion is not serious, and *"The beams at Stepney, according to Jan Bobrowski and Partners, had an average water-cement ratio of 0.35 and a maximum water-cement ratio of 0.38."* We are further reassured by a statement that *"At an age of six or eight hours minor changes in water-cement ratio reveal themselves in major changes in expected strength.*

It is therefore unlikely that variations in water content of the concrete leading to low minimum strengths after conversion could have occurred to any extent, though there must remain a small possibility that a minor increase in water content could have occurred at the top surface of the beams due to the finishing process."

The article suggests that conversion is a 'red herring', and, in the face of a previously published report[72], says *"At Camden it seems unlikely to have been of any consequence, as the same collapse mechanism of the beams falling off their inadequate bearings could have occurred with a Portland cement concrete.*

At Stepney, the use of high alumina cement is probably of more consequence as the cause is more likely to be related to some susceptibility of the material. However, if chemical attack is established as a major cause of the collapse, should the high alumina cement or the woodwool be blamed?"

This persistence in refusing to consider HAC as the root of the trouble is strange. As late as 9th May, 1974, the *New Civil Engineer* wrote: *"All that BRE knows for certain is that sulphate and other chemical attack did take place at Stepney, the beams as expected had converted and lost strength, but then something else has happened which reduced strengths as to as low as 3.5 N/sq mm. Stepney's pool beam investigation has all along been hampered by the decision on site, taken on the day of collapse, to dismantle the whole roof, a process which took until the next day and destroyed most of the evidence."*

A rather surprising statement says: *"Danger is much reduced if the beams are composite with a floor slab. It is only likely that odd beams are defective and in a composite floor these are not likely to precipitate a catastrophic collapse."* One wonders whether this sort of patchwork idea is seriously advocated as a rational approach to design.

Another comment in the *New Civil Engineer* (16th May, 1974) is disarming: *"At first sight, the Stepney failure seems the result of slightly defective beams being used in an aggressive environment that was not fully recognised."*

We are reassured that *"In other buildings gross deflections seem to be giving ample warning of defects, or would do if maintenance and inspection, the price that must be paid for buildings of low first cost, were up to scratch. In that sense, there is less cause for alarm than for the sudden 'brittle' failure at Camden.*

What is cause for concern, however, is the manifest need for regular inspection of structures using this material, by engineers with a level of competence and a knowledge of material behaviour that is greater than that usually devoted to routine inspections." It seems to be suggested that, instead of concentrating on a safe design we are to rely on regular inspection.

In the same issue of the *New Civil Engineer,* John Parkinson gives his explanation of the Stepney school failure: *"Production records again helped to explain another mystery. Only if the beams, normally demoulded after 16 hours, were kept in their moulds until they had reached maximum strength, could it be explained why excessive cambers did not give away the weaker beams. The weakness of the beams which fell into the swimming pool became credible when the records disclosed that they had been left in their moulds for some time after pouring, and Meteorological Office records showed that the weather on the day of casting had been wet."* And: *"This construction above a swimming pool presented a source of aggressive chemicals and physical effects that could combine to make life uncomfortable and short for high alumina cement concrete.*

Sulphates were available from the plaster and caustic alkali from the woodwool. The woodwool might also have contributed calcium chloride to the witches brew, which could have been the cause of corrosion of the woodwool edge channels.

Calcium chloride is a familiar cause of reinforcement corrosion in Portland cement concrete, but the prestresssing tendons in the Stepney beams were in perfect condition. This could have been because the chloride reacted with calcium aluminate in the high alumina cement and was removed from the system.

The only remaining anomaly to explain in the investigation so far although other factors could still emerge is the unreliable behaviour of the ultrasonic equipment."

Little of this was confirmed by the report[35] of the Building Research Establishment published the following month, although according to Parkinson's article of 4th July, 1974, *"the reasons for the failure given in the report are broadly in agreement"* with his article.

Some letters to the editor of the *New Civil Engineer* disagreed with Parkinson's attitude. On 18th July, 1974, A. Safier wrote: *"I was most surprised to read John Parkinson's article (New Civil Engineer 4th July) on high alumina cement concrete as the matter is much too serious to be handled in tabloid manner-journalism, giving part facts or even conjecture where precision is required in a professional journal.*

During the last year three collapses— not one, as reported by you—occurred in Britain due to, or contributed to by high alumina cement conversion. Camden and Stepney have been reported widely, the third, also in an educational building, has not.

Prof. Neville, in his paper 'A study of deterioration of structural concrete made with high alumina cement' (No. 6652 Proceedings of the Institution of Civil Engineers July 1963, discussion May 1964), records many more failures and clearly describes the unknown nature of the material and the difficulties in controlling its manufacture: of particular interest are the failures in Bavaria in 1961 reported in clause 93 of the paper and the action taken by the authorities there."

And further, *"To assume that the problem is restricted to X12 joists, or to any one manufacturer, is merely sweeping the problem under the carpet. High alumina cement conversion is simply unpredictable, dependent on so many factors that its performance cannot be guaranteed, and it would be illusory for the practising engineer to clutch at the straws of the tests.*

It is a travesty of the truth to say that educational buildings are 'light short life structures', or that they are poorly maintained. Indeed, generally, school buildings are better maintained than most. From my experience I can assure John Parkinson that consulting engineers do not assume that 'old age would show itself more gradually'—we design structures that will not suffer from 'increasing deflections and gaping cracks'. Witness the many ordinary Portland cement concrete structures still in use, built around the turn of the last century. The failures in high alumina cement concrete structures occurred within five or ten years of their construction and, if these are not replaced, more are likely."

But Parkinson answered that *"It seems illogical to ascribe the blame for a beam falling off inadequate bearings, as happened at Camden to the cement of which it was made"*, and wrote on 25th July, 1974: *"With memories of the aftermath of Ronan Point, which left an unnecessarily heavy millstone round the necks of designers, the British Standards Institution approach to advice for designers of future buildings is much to be preferred. After the Tay Bridge disaster, wrought iron was not banned, but the bad practice that contributed to the failure was recognised by engineers acting on their own responsibility. There seems now to be too ready a wish to usurp the engineer's responsibility judgement. How will the amendment propose that a designer cope with a foundation in acidic ground water? By tanking, which is every bit as workmanship sensitive as the proper use of high alumina cement."*

He added that *"Lafarge, maker of high alumina cement points out that the Building Research Establishment report on the Stepney collapse showed that the failure was caused by misuse of the material."*

Parkinson's own view is the *"all the failures of HAC in recent months have been specifically caused by misuse."* This sort of logic can be extended to explain virtually all structural failures by misuse: of analysis, of workmanship, etc.

In the *New Civil Engineer* of 1st August 1974, R.H. Elvery points out that the beams for the Stepney school *"had specified mix proportions of 1:3:6 by weight, which corresponds to a cement content of only about*

240 kg/m^3." He finds it therefore *"difficult to believe that a water-cement ratio of 0.40 could have been used consistently under production conditions without major difficulties arising in the compaction of concrete."* It appears that the water-cement ratio could not have been 0.40, and the cement content may be compared with the minimum of 400 kg/m^3 specified in France[80].

As late as 22nd August, 1974, John Parkinson wrote rather querulously in the *New Civil Engineer,* when commenting on the technical instruction of the Property Services Agency[94] : *"It is a pity that lapses on the part of the manufacturers of Portland Cement Concrete, resulting in excessive use of water, or calcium chloride, have not inspired the same opprobrium, especially as the Property Services Agency document mentions as a source of chemicals harmful to high alumina cement unspecified calcium chloride in adjacent Portland Cement Concrete."*

In the 16th January, 1975 issue of the *New Civil Engineer,* John Parkinson still thought that 'HAC troubles' appeared unexpectedly; he says that *"Lafarge, manufacturers of the cement, admit that they did not think of the possibility of existing structures built to the former confused advice being dangerous. It did not occur to anyone else either".* This is imputing ignorance to those who were in fact aware of the long-term behaviour of HAC. And we can add that the early advice was not so much confused as incorrect.

The Sunday Times

It seems thus that, while one part of the press can see nothing wrong in the structural use of HAC, another part grossly exaggerates the 'HAC problem' and indeed creates an 'HAC scare'. For instance, in reviewing a television programme (in which Mr Bobrowski, the City Engineer of Birmingham, and the Author appeared) the *Sunday Times* of 20th October, 1974 referred to HAC beams which *"turn to soil losing up to 99 per cent of their strength"* and felt that the implications of concrete with HAC were "*mind boggling."*

The true assessment of the 'HAC problem' clearly lies somewhere in between, and we shall try to present it in the remaining chapters.

12 Methods of Appraisal of Existing Structures.

12.1 Identification of HAC concrete and visual examination

In view of everything we have read in the preceding chapters it is clear that we must not simply ignore the presence of HAC concrete in our structures. We cannot say that it does not matter whether the concrete is made with HAC or with Portland cement, we cannot assume that a structure made with either material is equally strong and equally safe. The structural engineer has an excellent record of safety and he must guard it jealously.

It follows then that the safety of every structure containing HAC should be verified. But how do we know which structure was built with HAC concrete? In Chapter 9 we mentioned the advice of the Department of the Environment on the sources of information: the cement makers, the precast concrete manufacturers, the consulting engineers. Between them, they should identify the majority of buildings in which HAC concrete units have been used but not necessarily all, for in the case of a performance-specification approach the contractor was free to choose the actual material used. His records should show whether or not this was HAC but they may not be adequate, or indeed available, if the building is many years old.*

Visual inspection

Visual inspection may therefore be necessary to establish the presence of HAC in the concrete. This is not so simple, first because the concrete has to be uncovered. If we are dealing with floor units, their top part will usually be covered with a screed, topping or flooring, and their soffit will be hidden behind a false ceiling or plaster. The situation is similar in the case of a roof. It is therefore usually necessary to remove some of the finishes in order to have a look at the concrete units.

Here comes the second difficulty: it is not always possible simply to look at a concrete member and to say this is, or this is not, HAC concrete. Admittedly, HAC is darker than Portland cement but the actual colour of the concrete depends also on the colour of the aggregate and on its quantity. In practice, in the majority of cases, it is possible to identify HAC by examining the colour of the mortar matrix in the interior of the concrete. In constrast to

* There is virtually no way of discovering odd HAC concrete beams which may have been used in an otherwise Portland cement concrete construction. The reason for such an occurrence might have been an urgent need for a few extra beams during the construction and HAC allowed a more rapid production (see page 171). In Germany, many producers of precast beams used HAC (when its use was permitted) on Monday to Thursday and Portland cement on Friday. The latter was cheaper and the transfer of prestress would not be carried out till after the weekend anyway.

the 'cement colour' of Portland cement (light grey to fawn), HAC of British manufacture is dark grey to black or, if highly converted, often grey-brown or chocolate in colour. However, some other cements, such as sulphate-resisting Portland cement, are also dark, and certain aggregates, especially their fine fraction, can lead to a very dark appearance of the mortar matrix. Thus differentiation between HAC and Portland cement on the basis of colour is not always possible and recourse may be necessary to chemical tests. One of these is the differential thermal analysis test described in Section 12.2, but the simplest test is that developed by the Building Research Establishment[103].

The BRE test

This test is relatively easy and can be performed in less than 10 minutes. The underlying principle is that HAC concrete will yield large amounts of aluminium when treated with dilute sodium hydroxide, while Portland cement concrete will not. A sample of powdered concrete is obtained by drilling several holes with a masonry drill or by crushing a small piece of concrete with a pestle and mortar, picking out any aggregate, and grinding the remainder to powder. About 1 gram of the powder is then treated with appropriate reagents: abundant formation of a yellow precipitate indicates the presence of HAC; a clear, or slightly cloudy, solution indicates the presence of Portland cement, blastfurnace slag or pulverised-fuel ash.

An even simpler test, used in Germany, is to spray a phenolphthalein or thymolphthalein indicator on freshly broken concrete and to observe colour change.

Condition of the concrete

Having established that our building contains HAC concrete, we should next determine its condition and strength in relation to the design or service requirements. This is a major part of the decision as to whether the building is safe, and the task is not an easy one.

The various guidelines of the government departments were reviewed in Chapter 9 but some of the subject matter should be treated more fully, and this we propose to do now.

What to look for

The first step to take is to carry out a visual examination of the part of the structure made of, or supported by, HAC concrete. In the majority of cases, there will be no signs of change in the external appearance of the concrete, but it is still useful to have a careful look. Examples of what we are looking for are: a large deflection, a pool of water on the roof, lateral bowing, cracking or discoloration of concrete. The last one, whatever the colour, *may* indicate a chemical attack of HAC. If a part of the HAC concrete looks moist, be it due to leakage or to condensation, this is the place to be examined with care. Abnormal deflection is well worth looking for but this may be masked by a false ceiling, whose state can be very misleading. (On one occasion, a deflection of 200 mm in a roof span of 9.7m was discovered after the removal of a false ceiling not connected to the roof beams.) We should note that the deflection of beams with infill pots and with a screed is only about one-third of the deflection of similar beams in isolation. Even without the screed but with infill pots, the deflection is reduced by one-half compared with isolated beams of the same size.

Any soft concrete, or any of the signs of structural distress mentioned above, means that all is not well and that remedial measures have to be taken. However, in the majority of cases these gross signs of trouble will be absent, and we have to do further tests.

It may be convenient at this stage to quote from the Property Services Agency Technical Instruction CE 100 B62 issued in December 1974[100]. This lists *"Points to be considered when making an engineering appraisal"* as follows:

"1. Date of construction/age of building.

2. Type of construction in roof/floors:
- *i. Composite*
- *ii. Non composite*
- *iii. Structural screed to roof/floor provided*
- *iv. Beams designed as Tee beams incorporated in structural slab*

See para 16 of the Technical Instruction

3. Loading:
- *i. Dead*
- *ii. Live*

Is this less than design load or can it be reduced?

4. Stresses: Was design based upon fully converted HAC concrete strengths or a reduced strength. Whenever possible assess residual factor of safety by checking working stresses against core crushing strength.

5. Leased buildings:
- *i. Is it a full repair lease?*
- *ii. Length of unexpired lease*

6. Occupational use of buildings:
- *i. Previous*
- *ii. Present*
- *iii. Future*

Changes:
a. Use of building
b. Ventilation system in its operation, may increase humidity or heat produced.

7. Environmental conditions:
- *i. Temperature inside building*
- *ii. Humidity inside building*
- *iii. Likelihood of roof leaks, condensation*
- *iv. Localised areas of heat (see para 12)**

8. Building materials adjacent to HAC member
Are there any materials which contain free salts of:

† *"16 Composite construction as used in this Technical Instruction means that the structural members have been designed to act in conjunction with a reinforced structural topping with shear links. Note particularly that a mesh reinforced sand/cement screed should not be regarded as composite construction ..... It should be noted that HAC concrete joist hollow pot floors or roofs are regarded as non composite construction but are in a lower risk category than isolated non composite beams."*

* *"12 The immediate cause for concern is for buildings or parts thereof incorporating long spans with non composite isolated beams in which conversion may be well advanced. Some structures are likely to be more vulnerable than others, e.g. Swimming Pools, Kitchens, Boiler Rooms, Plant Rooms, etc. where high temperatures and humidity are more likely. Particular attention should be paid to the presence of localised sources of heat such as light fittings, heating panels, hot water pipes, etc. as local effects on structural members may be caused by these, reducing the strength of the member concerned."*

	Possible sources
Alkalis	*Gypsum plaster contains sulphates*
Chlorides	*Woodwool contains alkali and calcium chlorides*
Sulphates	*Portland cement—caustic alkalis and possibly unspecified calcium chloride*

9. *Characteristics of concrete:*
   - *i. Mix proportions*
   - *ii. Water-cement ratio* — *Actual w/c may have exceeded design w/c during manufacture*
   - *iii. Aggregate* — *Granite or gravel containing feldspar and micas produce free salts of sodium and potassium*

10. *History of HAC members:*
    - *i. Name of manufacturer*
    - *ii. Date of manufacture*
    - *iii. Place of manufacture*
    - *iv. Details of curing*
    - *v. Members cast outdoors/under cover."*

12.2 Differential thermal analysis

Degree of conversion

Following the visual examination, the next step, recommended by the Department of the Environment[92], is the determination of the degree of conversion of HAC in the concrete. We defined the degree of conversion in Section 3.3. The test result tells us what proportion of the hydrated aluminates has changed from hexagonal to cubic form. The degree of conversion is determined by a sophisticated chemical test known as differential thermal analysis, or briefly DTA, performed on a small sample of hydrated cement or mortar.

The underlying principle is that for each compound there is a dissociation temperature at which a large amount of heat is released. This shows as a peak on a thermogram, i.e. a plot of the temperature of the sample versus the ambient temperature. Thus a quantitative as well as a qualititative analysis of the various compounds is possible. In our case, we are interested in the hexagonal and cubic calcium aluminate hydrates (see Section 1.2). The procedure is to heat a sample of a few grams of hydrated HAC with a similar sample of inert material, usually alumina, in a laboratory furnace. The rise in the temperature of the inert material is measured by a thermocouple and is controlled to be as uniform as possible. The temperature of the HAC sample is also determined, and the difference between the two which occurs at specified temperatures, and shows as a peak, indicates the presence and quantity of the various compounds. Specifically, $CaO.Al_2O_3.10H_2O$ dissociates at around 100°C and $3CaO.Al_2O_3.6H_2O$ at 300°C to 330°C. The test identifies also any chemical attack of the concrete in the area which has been sampled.

There are some practical difficulties in applying the DTA technique to HAC. The presence of water in some of the compounds in hydrated HAC masks the sharpness of the temperature peaks because dissociation takes place over a temperature range rather than at a specific value. In such a case, the results may not be easy to interpret.

The second difficulty arises from the possibility of accidental conversion of HAC in the sample. Before discussing this further, we should explain how the sample is obtained in the first instance. The advice given by the Department of the Environment[92] is as follows: "*Samples may be obtained by drilling small holes, for example with a No. 8 masonry drill. Care should be taken to ensure that all plaster or other finishing materials have been removed before drilling is commenced. It is suggested that a hole should be drilled into the concrete to a depth of about $\frac{1}{4}$ in. (5 mm) then blown clear to remove any surface particles. The hole should be further drilled for about $\frac{3}{4}$ in. (20 mm) and the drillings caught on a card. It is suggested that four such holes be drilled over a distance of 1-2 ft (300-600 mm), and the mixture of the drillings from the four holes be placed in a sealed container and sent for analysis. Beam samples should be collected from regions both at midspan and near to a support. Initially such samples should be taken from at least two or 10 per cent of the members which should be clearly classified as to position in the structure and age in years. If there are any obvious areas of chemical attack, samples should be taken from the attacked zone, and also from areas subjected to exceptional environmental conditions.*"

We are told further: "*It should be noted that this test, as well as establishing the degree of conversion, will also identify chemical attack on the high alumina concrete, if it occurs at the sampling points.*

Where the degree of conversion of the high alumina concrete varies by more than 10 per cent, further members should be tested."

Suggestions have been made that some samples yield unduly high values of conversion because of the high temperature induced by drilling[101]. This may occur but can certainly be readily avoided: blunt drilling bits should not be used, and when the bit hits a large aggregate particle a new hole should be drilled. Alternatively, a sample can be obtained by chipping a piece of concrete with a hammer, then removing a part remote from the outer surface and grinding without heating.

We may also note that significant amounts of clay minerals or mica in the aggregate affect the results of DTA; it is possible to minimize these effects by removing as much of the aggregate as can be identified before submitting the sample to DTA.

These problems have led the Building Research Establishment to issue a press release[102]. This says the following: "*Possible shortcomings of the method, which have been widely quoted, are:*

A *Over-heating of the drill may increase the degree of conversion of the sample.*
BRE, Local Authorities, and Testing Houses have all carried out experiments in which results from samples taken by drilling are compared with those from crushed specimens from the interior of beams. The experiments covered degrees of conversion ranging from 20 per cent to 90 per cent and demonstrated that the degree of conversion was increased only to a negligible extent by the drilling method, provided

that ordinary good drilling procedures were observed, for example the avoidance of an extremely blunt drill. BRE is satisfied that using these procedures overheating leads to no significant error.

B If the drill strikes aggregate, dilution of the specimen may lead to error. With modern DTA equipment this will not happen. The DTA technique is based on a comparison of certain peaks in the thermogram. Dilution of the samples reduces the heights of the peaks but does not affect their ratio on which the test depends.

C It might be better to employ samples chipped from the structural members.
It is not. The degree of conversion of the outer layer can be much less than that of the internal concrete. (In the beams in the Sir John Cass School, Stepney, the internal concrete was 85 per cent converted; that in the outer layers only 40 per cent.) It follows that the use of chipping to obtain samples, unless these are so large as to impair the structure, involves serious risk of under-estimating the degree of of conversion and thus of over-estimating the quality of the concrete.

Thus the Building Research Establishment view is that these alleged shortcomings do not affect the value of the test in preliminary diagnosis."

With reference to the last point we can note that tests at Birmingham have shown good agreement between the results of the DTA on drilled and chipped samples.

The press release of the Building Research Establishment does not deal with one other criticism of the DTA test which has been levelled at it: that the test results are not reproducible. This may be so with laboratory operators who lack skill and experience, and unfortunately in the rush to deal with the 'HAC problem' some inexperienced operators have been involved. It should also be admitted that trace elements may affect the results, and in Germany X-ray diffractometry is preferred; this is, however, more expensive than DTA. We can accept that for practical purposes a competent laboratory will obtain adequately reproducible results of the DTA tests. Thus, as far as the determination of the degree of conversion is concerned, the test is good, but it is not at all obvious how one interprets the test result for the purpose of estimating the strength of the structure. We shall refer to this in Sections 13.1 and 14.3 but at this stage let us look at the other tests available to the engineer.

Capillary test

One of these is mentioned in the guidelines of the Institution of Structural Engineers [104], and is recommended by the Lafarge Aluminous Cement Co. Ltd. The test determines the capillary rise of water in a sawn sample of HAC concrete placed on a piece of wet felt. The height of the rise is claimed to be related to the degree of conversion in a very simple manner. There is a proviso that the relations apply to *"good quality, well compacted concrete (e.g. 0.4 water-cement ratio)"*[104], but of course we never know what was the water-cement ratio of the concrete which is being tested. Indeed, if we could be *certain* that the water-cement ratio was not greater than 0.4, we probably would have a reasonably safe structure. Moreover, the relation between the capillary rise and the degree of conversion, as given, is so simple and so crude that one finds it difficult to accept it without some proof. This has not yet been offered. We fear therefore that the test is not likely to yield any reliable and useful information.

12.3 Core test

A more direct test is to determine the strength of cores (i.e. small cylindrical samples) cut from the HAC concrete members. This was mentioned in Chapter 9 but there are some serious practical problems in core cutting. These arise from the difficulty of access to many HAC concrete units, already mentioned, and the difficulty of drilling in any direction other than vertically down. It is also important not to cut the prestressing steel (other than the top wire whose function is to prevent tensile stresses in the member prior to its being put into service), and indeed not to damage the member structurally: the core test is not entirely non-destructive. The greatest difficulty arises from the fact that the structural members involved are generally too small to permit the cutting of standard cores (as prescribed by the British Standard BS 1881: Part 4: 1970) with a diameter of 100 mm. We should note that a desirable length of a core is twice its diameter, and a minimum length is equal to the diameter. In practice, even 75 mm diameter cores can rarely be drilled, and it may be necessary to use a diameter as small as 25 mm. Cores of this size may yield results which are of little use and, for instance, the Property Services Agency[94] requires a minimum diameter of 75 mm and a minimum length of 100 mm.

The testing of cores is prescribed by the British Standard BS 1881: Part 4: 1970 but the interpretation of results in relation to the strength of concrete in the structure is not simple. The main reason for this is that the strengths of concrete specimens made of the same concrete but differing in size and in shape (cubes and cylinders with different height-diameter ratios) are not the same[9]. Thus numerical values obtained for a particular set of cores cannot be simply compared with the strength of standard cubes which was originally specified and used in design. Some conversion factors are given in the British Standard BS 1881: Part 4: 1970 but these are inadequate for the purpose[9].

Furthermore, cores often show a considerable scatter in values, which are affected by local compaction, bleeding (the bottom of a member being often stronger than the top, but this applies to the as-cast and not necessarily as-used position), cracking, or presence of reinforcement, and indeed by the process of drilling. The curing history of the concrete from which the cores were cut is also likely to be different from that of the standard test cubes.

Nevertheless, the core test is a useful one in that it gives a broad indication of the quality of the concrete, its compaction, distribution of aggregate and, above all, of the strength of the concrete in the given members. If the core strengths correspond to equivalent cube strengths around 40 N/mm^2, or even 30 N/mm^2, then the concrete, whether it has undergone conversion or not, is probably strong enough for structural purposes. On the other hand, if the core tests yield values of 15 N/mm^2 or 20 N/mm^2 then the concrete has a poor strength and is unlikely to be good enough for service in prestressed concrete members. We feel therefore that the core test, if possible and if carefully executed, is useful in the preliminary assessment of a structure.

12.4 Ultrasonic pulse velocity test

We have earlier expressed the view that the core test is not entirely non-destructive although, as a rule, it is possible to make good the damage caused by the drilling of the cores. There exist, however, some tests which are entirely non-destructive. The most prominent among these is the ultrasonic pulse velocity test. Here, two transducers are placed on opposite sides of a piece of concrete, a known distance apart, and the time for an ultrasonic pulse to travel from one transducer to the other is measured. This time, divided by the distance travelled, gives the pulse velocity; hence the name ultrasonic pulse velocity. We should stress that this is all that is measured: any translation of the velocity into strength is possible only if we know the various mechanical properties of concrete (which we do not). In practice, we can translate the pulse velocity into strength only when we are dealing with a given concrete mix, i.e. the type of aggregate and the aggregate-cement ratio must remain the same. Moreover, the translation must be based on a calibration of the ultrasonic pulse velocity on cubes whose strength is actually determined.

Thus the ultrasonic pulse velocity does not tell us the strength of the concrete but makes it possible to compare the strength of similar concrete units. The reason for this is that the pulse velocity is affected by the density of the material through which the pulse passes. Specifically, the velocity through air is low so that the presence of cracks and pores decreases the velocity. Thus, highly converted HAC concrete would register a lower ultrasonic pulse velocity than a similar concrete unconverted, and cracks due to structural distress would also lower the velocity.

Difficulties of the test

But there are some serious difficulties: cracks perpendicular or slightly inclined to the line along which the wave is sent reduce significantly the pulse velocity. However, cracks parallel to the path of the pulse do not affect the velocity to a great extent as the pulse can pass through the adjacent concrete. Because it causes dilation, a tensile stress in the relevant direction reduces the pulse velocity, while a moderate compressive stress may increase it. The presence of reinforcing or prestressing steel affects the velocity, especially when the pulse can travel along the steel. The moisture condition of the concrete also influences the pulse velocity, which is higher through water than through air.

A further, and major, criticism of the ultrasonic pulse velocity test is that it is relatively insensitive to changes in the strength of concrete (cf Fig. 12.1). In some cases, it was found that the pulse velocity through a precast concrete member was higher after conversion (induced by heating the member) than before conversion. In other cases, beams with a higher pulse velocity had a lower ultimate strength than the beams with a lower ultrasonic pulse velocity measurement. This type of behaviour is confirmed by recent laboratory tests at the University of Nottingham [112]. The tests were on HAC concrete (unlike all previous work, which was on Portland cement concrete) with both continuously graded and gap-graded aggregate. Varying water-cement ratios and different curing conditions were used. The results of the investigation cannot be described as conclusive but in some cases a strength variation of 40 to 50 per cent was accompanied by no difference in the pulse velocity. We must stress, however, that the investigation was on laboratory specimens, with conversion induced rapidly rather than taking place over a period of years.

An investigation[113] published at the same time as the Nottingham paper[112] reports that the relation between the pulse velocity and the strength of concrete is not well defined in the case of concretes of very high strength. Since the initial strength of HAC concrete falls into this category, the reduction in strength on conversion is accompanied by only a small reduction in pulse velocity. This reduction is often insufficient to give a clear indication of the strength of concrete in an unstressed state. The reasons for this are not entirely clear but this is not of importance; what is, is that the test is not sensitive enough to detect a loss of strength on conversion and, after all, this is what we are seeking.

The paper[113] identifies a further source of error peculiar to concrete under high stress: such a stress may induce microcracking in a direction normal to it. As a result, if the pulse path is normal to the principal compressive stress in excess of 35 per cent of the ultimate compressive strength, the pulse velocity is reduced compared with that in other directions.

Since the ultrasonic pulse velocity test is relatively easy to perform and therefore often viewed with favour, it is useful to consider the subject further and briefly to review an investigation of the ultrasonic pulse velocity of HAC concrete in service[114]. Cores were cut out from the members and an attempt was made to correlate the core strengths and the pulse velocity. Two types of concrete were involved, one made with 10 mm crushed granite, the other with 20 mm gravel. For the former, a good correlation was obtained but in the case of gravel concrete, strengths ranging from 30 N/mm^2 to 80 N/mm^2 all corresponded to a pulse velocity of about 4.4 km/s. Numerous factors were thought to account for this, the main being the small size of the cores used (43 mm in diameter) and therefore some uncertainty about their strength. It also seems that the use of ultrasonic pulse velocity on in situ members of small cross-section is really difficult: if we test across the web, the path length is very short and difficult to measure with sufficient accuracy; if we measure across the flange, the path is longer but the presence of the steel affects the reading; moreover, the dimension at right angles to the pulse direction may be too small, a minimum of about 80 mm being recommended for a pulse frequency of 50 kHz. We believe that in the investigation in question[114], the relatively large size of the gravel aggregate may have increased the variability of the pulse results.

Some of the data of the Nottingham investigation[114], as well as those of the Building Research Establishment[35], are shown in Fig. 12.1. We conclude from this figure that to use the pulse velocity as an indication of the strength of concrete would be extremely unwise.

To justify this view further we may look more closely at the experiences in the Stepney school investigation (see Chapter 7). The Building Research Establishment[35] took a large number of ultrasonic pulse velocity readings; some of these were on two beams from the gymnasium roof, the pulse velocity being determined across the top flange of the beams in situ and later in the laboratory. Cores were taken from the same part of the beam, and the pulse velocity was determined on the cores. The compressive strength of the cores was then determined, presumably to establish a relation between the pulse velocity and strength. Some of these data are given in Table 7.2.

A statistical analysis[115] of these data was made by someone from outside of the Building Research Establishment. This analysis showed a very high scatter of the pulse velocity readings and suggested that there was no correlation between the pulse velocity and the strength of concrete in a given member. The same paper[115] made also some general points which are of interest:

1. The thickness of concrete used to determine the ultrasonic pulse velocity should be not less than 100 mm when the maximum aggregate size is 20 mm or more;
2. The path length of the pulse should be determined within ±1 per cent;
3. The acoustic coupling between the surface of the concrete and the transducer must be good; in some cases, this may mean that the surface of the concrete has to be ground or made good.
4. The influence of the reinforcing steel on the pulse velocity has to be borne in mind.

Figure 12.1 **Relationship between ultrasonic pulse velocity and equivalent cube strength.**

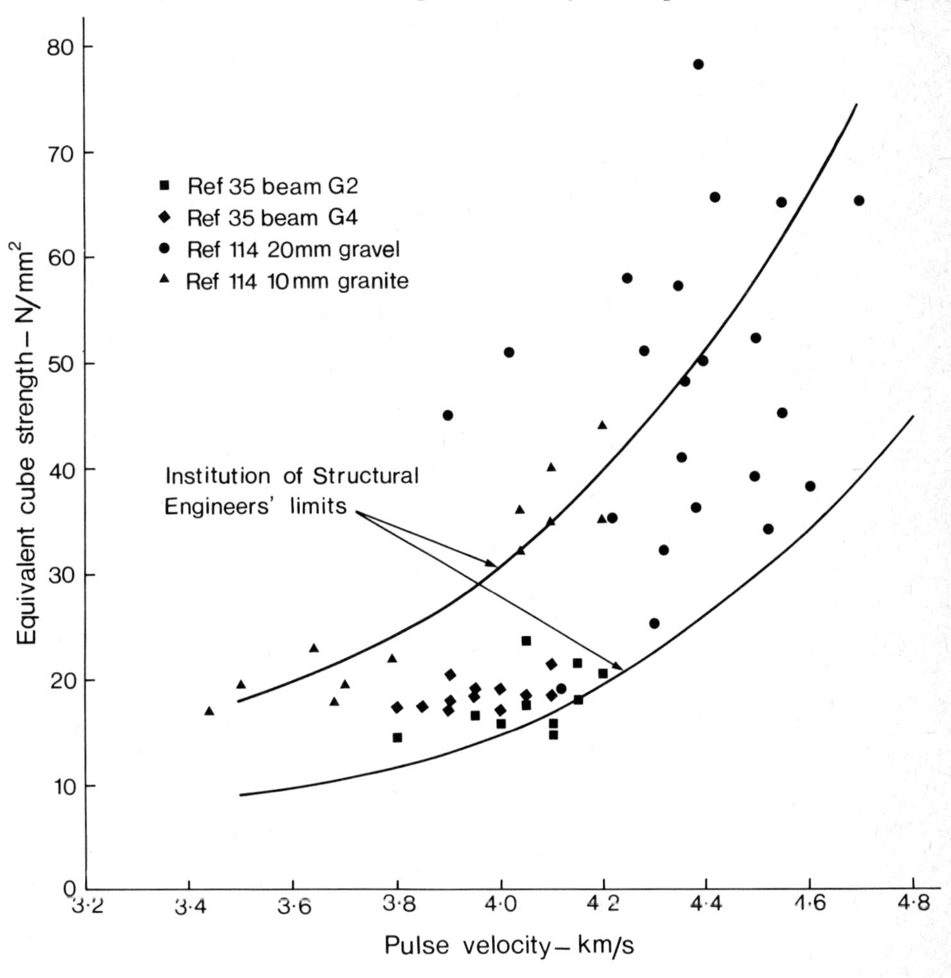

In view of all this, we feel that the advice of the Department of the Environment[92] is rather optimistic when we are told that *"Estimation of strength can only be made by cutting cores and obtaining a relationship between pulse velocity and strength for a particular set of concrete cores."*

We do not agree either with the guidelines[104] of the Institution of Structural Engineers, which say that when the working stress is close to the crushing strength of the concrete, the test *"provides a realistic warning that the safety margin is low"*. The report also contains a graph (reproduced here as Fig. 12.1) which shows *"the probable relation between the ultrasonic pulse velocity values and the* ***cube*** *strength of HAC concrete"*. The word 'probable' should be noted. Anyway, the range of the data is such that we would find it difficult to interpret the results. For instance, at a pulse velocity of 4.2 km/s, the cube strength, as given by the graph, ranges from 40 N/mm^2 (which is very satisfactory) down to 20 N/mm^2 (which is highly worrying). However, if the pulse velocity is below about 4.0 km/s, this can be taken as an indication of the presence of advanced microcracking and probably means that the actual stresses are approaching the crushing strength of the concrete.

We would like to say again that we do not consider the ultrasonic pulse velocity to be of use in assessing the strength of HAC concrete in a structure. Nevertheless, the test is of some value in conjunction with cores. Although we do not agree with the circular[92] of the Department of the Environment, which says that estimation of strength can be made by the use of both tests combined (see earlier), we feel that the ultrasonic pulse velocity is of help in seeking out areas of weakness. We doubt that more than that can be expected.

Finally, it may be appropriate to mention that the ultrasonic pulse velocity test, although it seems simple, requires an experienced operator. In view of the large number of tests currently being carried out, we suspect that some of the people involved lack this experience. This may lead to unreliable data or, at best, to a high scatter of results. If we then try to correlate the ultrasonic pulse velocity results with those of core tests, we find ourselves in a position where the errors in the two tests can be so large as to render the final answer—the strength of the concrete in the structure—highly unreliable.

If, for any reason, ultrasonic pulse velocity measurements are taken on a HAC concrete structure, the number of readings must be adequate to be of any value. For instance, the Property Services Agency[94] recommends not fewer than five readings per member and not less than 20 per cent of the members to be tested.

12.5 Load test

There is one more non-destructive test available to the engineer who wishes to estimate the strength of a structure: the load test. This test can be used as an alternative to the preceding tests or can be employed in addition to them. The load test essentially measures the performance of a part of a structure under a known superimposed load. The procedure is standardised in the Code of Practice CP 110:1972[86] and falls into two parts. The first one concerns load tests on individual precast units. The code says: "*The load tests described in this clause are intended as checks on the quality of the units, and should not be used as a substitute for normal design procedures. Where members require special testing, such special testing procedures should be in accordance with the specification.*

The procedure

Test loads are to be applied and removed incrementally.

The unit should be supported at its designed points of support and loaded for five minutes with a load equal to the sum of the characteristic dead load plus $1\frac{1}{4}$ times the characteristic imposed load. The deflection should then be recorded. The maximum deflection measured after application of the load shall be in accordance with the requirements which should be defined by the engineer.

The recovery should be measured five minutes after the removal of the applied load and the load then re-imposed. The percentage recovery after the second loading must be not less than that after the first loading nor less than

90 per cent of the deflection recorded during the second loading. At no time during the test should the unit show any sign of weakness or faulty construction as defined by the engineer in the light of a reasonable interpretation of relevant data."

A destructive test on individual units is also presented. We are told: *"The unit should be loaded while supported at its design points of support and must not fail at its ultimate design load within 15 minutes of the time when the test load becomes operative. A deflection exceeding 1/40 of the span is regarded as failure of the unit."*

Another clause of the code[86] prescribes load tests of whole structures or of parts of structures. As in the case of individual units, the loads are to be applied and removed incrementally. The following clauses are quoted in full:

"9.6.3 Test loads. The Test loads to be applied for the limits states of deflection and local damage are the appropriate design loads, i.e. the characteristic dead and imposed loads. When the ultimate limit state is being considered, the test load should be equal to the sum of the characteristic dead load plus 1¼ times the characteristic imposed load and should be maintained for a period of 24 hours. If any of the final dead load is not in position on the structure, compensating loads should be added as necessary.

During the tests, struts and bracing strong enough to support the whole load should be placed in position, leaving a gap under the members to be tested, and adequate precautions should be taken to safeguard persons in the vicinity of the structure.

9.6.4. Measurements during the tests. Measurements of deflection and crack width should be taken immediately after the application of load and, in the case of the 24 hours sustained load test, at the end of the 24 hours loaded period, after removal of the load and after the 24 hours recovery period. Sufficient measurements should be taken to enable side effects to be taken into account. Temperature and weather conditions should be recorded during the test.

9.6.5. Assessment of results. In assessing the serviceability of a structure or part of a structure following a loading test, the possible effects of variation in temperature and humidity during the period of the test should be considered.

The following requirements should be met:

(1) For reinforced concrete structures and Class 3 prestressed concrete structures the maximum width of any crack measured immediately on application of the test load for local damage should not be more than two thirds of the value for the limit state requirement (see [2.2.3.2]). For Class 1 and Class 2 prestressed concrete structures, no visible cracks should occur under the test load for local damage.

(2) For members spanning between two supports, the deflection measured immediately after application of the test load for deflection is to be not more than 1/500 of the effective span. Limits should be agreed before testing cantilevered portions of structures.

(3) If the maximum deflection in millimetres shown during a 24 hours under load is less than $40l^2/h$ *where* l *is the effective span in metres and* h *the overall depth of the construction in millimetres, it is not necessary for the*

recovery to be measured and requirements (4) and (5) do not apply.

(4) If, within 24 hours of the removal of the test load for the ultimate limit state as calculated in 9.6.3, a reinforced concrete or Class 3 prestressed concrete structure does not show a recovery of at least 75 per cent of the maximum deflection shown during the 24 hours under load the loading should be repeated. The structure should be considered to have failed to pass the test if the recovery after the second loading is not at least 75 per cent of the maximum deflection shown during the second loading.

(5) If, within 24 hours of the removal of the test load for the ultimate limit state as calculated in 9.6.3, a Class 1 or Class 2 prestressed concrete structure does not show a recovery of at least 85 per cent of the maximum deflection shown during the 24 hours under load, the loading should be repeated. The structure should be considered to have failed to pass the test if the recovery after the second loading is not at least 85 per cent of the maximum deflection shown during the second loading."

Some of the terms and requirements can be understood only by reference to the actual code but in the vast majority of precast prestressed concrete units made with HAC we are concerned with Class 1 prestressed concrete.

We should note that the test load recommended consists of the actual dead load plus 1.25 times the characteristic imposed load, which is the load given in British Standard Code of Practice CP 3: Chapter V, Part 1:1967 Code of basic data for the design of buildings: Loading. For instance, for school rooms the distributed live load is 3.0 kN/m^2 and for flats, 1.5 kN/m^2. The Department of the Environment[92] suggests a somewhat higher test load, viz. 1.25 times the dead load as well as 1.25 times the live load. The reason for this is that we are dealing with *"a material which can lose strength with time"*. Also, there is probably a greater inherent uncertainty about HAC concrete and the variability of HAC concrete units. Furthermore, because of the possibility of future reductions in the strength of HAC concrete, the Department of the Environment[92] recommends that the load test be repeated in the future and that a more onerous load be applied.

The circular[92] of the Department of the Environment contains a warning which should be heeded: as in the case of Portland cement concrete, a very high sustained load reduces the ultimate strength of the concrete. For this reason, even if a concrete member made with HAC is strong enough now, but the stress actually acting is at least three-quarters of the ultimate strength, deterioration should be expected so that failure may occur without further conversion or its effects.

The load test cannot, as a rule, test the concrete member, or part of a structure, in flexure, shear and bond simultaneously to the same extent. We assume, however, that any deterioration of the concrete affects all its properties to the same extent. This was considered in Section 4.1. We should note, however, that after conversion the transmission length necessary to develop bond is increased.

Value of the test

The load test, at first sight, gives the impression of being a convenient and direct method of assessing the strength of a structure. The test is, however, not popular with some engineers. First of all, it is not easy to perform as safety scaffolding has to be inserted and large loads are involved; these may be in the form of bricks or bags of cement but containers filled with water are probably least inconvenient (Plate 14.5). There is also a difficulty of loading a sufficiently large area of the structure. The problem is aggravated by the uncertainty about the extent of composite action when this is not fully structually assured. In the case of roofs, the design live load is low so that the load test is not onerous. For these reasons, the load test alone is not considered a satisfactory means of determining the state of health of a structure. This is, for instance, the attitude of the Property Services Agency[94].

12.6 Destructive tests

Let us now consider a more direct, and destructive, test on the strength of the actual concrete members, reinforced or prestressed. This necessitates the removal of structural components, undamaged if possible, and testing them to destruction in the laboratory. The test is usually in flexure, although occasionally the strength in shear may be determined. Such a test gives a reliable value of the strength of the member tested but only a guidance to the strength of similar members. Specifically, the test gives no information on the strength of other members which may have suffered damage from local defects, from unfavourable environmental conditions, or from bad workmanship in casting.

According to the guidelines[104] of the Institution of Structural Engineers, a load factor of 1.4 on the dead plus live load would provide an acceptable margin of safety except for assembly halls and the like where a higher value should be expected. The value of 1.4 is lower than the value specified in design by the Code of Practice CP 110:1972[86], but the guidelines[104] point out that *"As with every existing structure, the assessment (of stability) will generally require a different approach to that used in the design of a new building in that, once it is built, many of the 'uncertainties' have been determined. The essential requirement is then to ensure that there is a reasonable margin of safety to protect the user and it is usually unnecessary for an existing structure to require the same theoretical factor of safety as for a new design "*

We should make it clear that the calculation of the load factor need not be based on the actual ultimate strength of a member but can be made by reference to a calculated strength of the concrete in the member. In such a case, the strength of concrete should be taken as 0.67 times the assessed cube strength.

We may mention the German approach, which allows for future changes by requiring a load factor of 1.5, using as concrete strength 28 per cent of the original one-day strength (usually 60 N/mm^2). But even with this load factor satisfied, HAC concrete has to be permanently propped.

13. Safeguards for the Future

13.1 Monitoring the safety

We have described various tests but carefully avoided answering the question, which test is best? The trouble is that no test can directly tell us how safe a structure is. What is necessary in addition to the values of the degree of conversion, strength of concrete cores, ultimate strength of members, or the deflection in the load test—in any combination, or even all of these—is the engineer's judgement. As the guidelines[104] of the Institution of Structural Engineers say, *"Many building owners will expect a definitive statement on the condition and safety of their buildings. The situation regarding HAC concrete as outlined in these Guidelines should be explained by the Engineer as far as possible. It should be emphasized that, at least for the immediate future, only engineering judgements can be offered."*

Future behaviour

Difficult as a reliable assessment of the current strength is, the prediction of future behaviour and development of strength is doubly difficult. Perhaps the easiest case is when the degree of conversion is very high: the concrete is unlikely to lose any more strength unless it is subjected to chemical attack. This can be guarded against but is still possible, for instance, when a roof leaks through alkali-bearing material or when aggressive washing liquids find their way to the HAC concrete.

Because of the uncertainty about the future, and also because of the doubts surrounding all the tests, there is a strong case for continuing monitoring of the behaviour of the structure. In essence, this means that we measure continually some feature of the structure whose change would give a warning of impending failure or at least would alert us to a change in circumstances. This may sound vague but the logic is that no news is good news, and a change tells us that the structure, or the member, is moving nearer to collapse.

Methods of monitoring

Several methods of monitoring are currently in use. One of these is a system[105] which uses repeated ultrasonic pulse velocity measurements. Several transducers are permanently fixed to the beam, as shown for example in Fig. 13.1. Pulses are sent between the various transducers so that a large number of pulse velocity determinations in various directions is made. This gives information about the state of virtually all the concrete in the beam. The method also makes it possible to pinpoint a suspect area when a low reading in one direction is obtained; pulse velocities of intersecting paths are then determined.

Figure 13.1 **Typical layout of ultrasonic transducers used in a permanent monitoring system.**

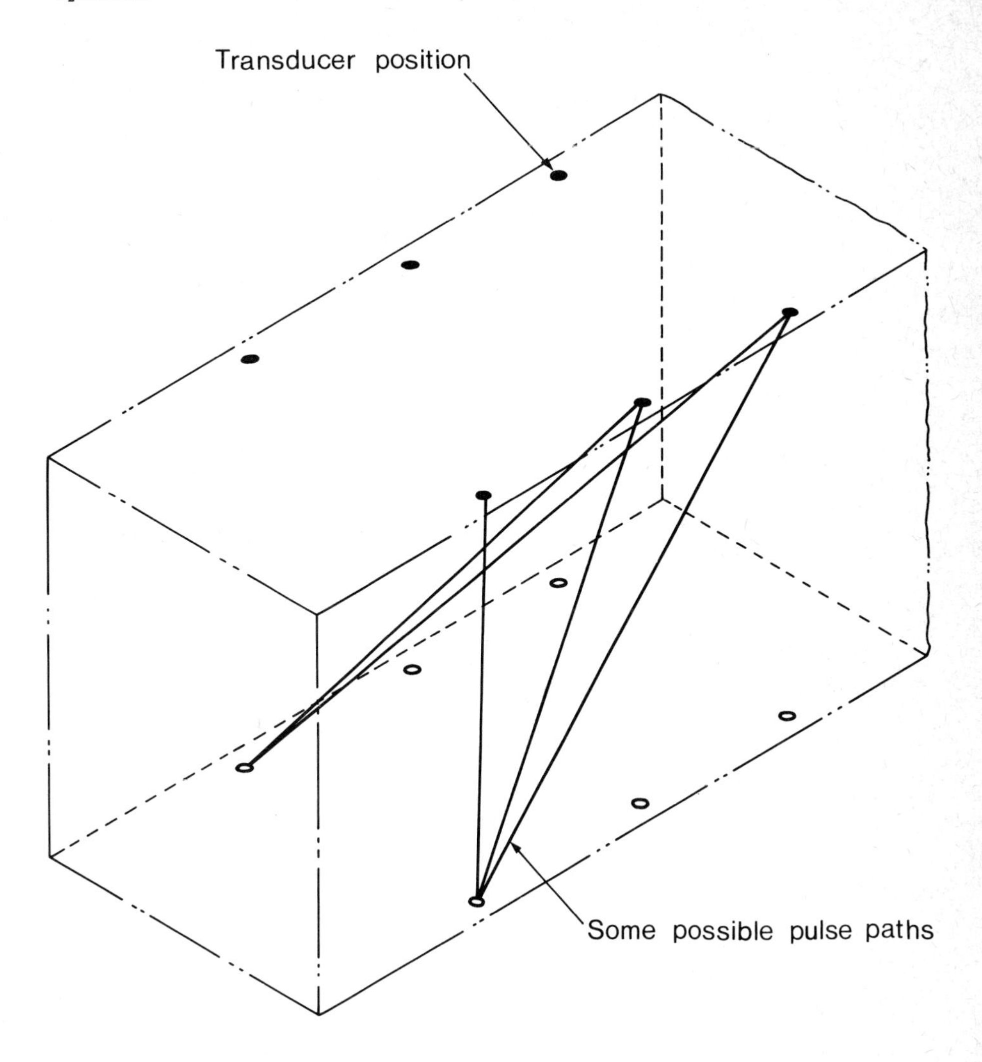

The system makes no attempt to evaluate the strength of the concrete but only alerts us to changes in pulse velocity and therefore, presumably, in conversion. The approach is fairly reliable because the installation is permanent so that errors due to acoustic coupling, path length measurement, and presence of reinforcement are eliminated. It is recommended that initially readings be taken daily but, when reproducible values have been obtained, the frequency can be reduced to once a month or even once every three months. Automatic recording is of course possible.

The system is unfortunately expensive. At the time of writing this book, the cost of supplying and installing each test point is about £23. It is recommended to install between 10 and 16 test points in a 10m length of a beam. With the lower number, there are 16 paths for which readings can be obtained; with 16 test points, there are 22 readings. The cost of taking readings has of course to be added to the initial cost. We are thus talking about a fairly considerable expense of monitoring just one beam, and this clearly is not enough in any building. The cost per building is therefore likely to run into several thousands of pounds. Of course, this may be money well spent if, as a result, the building is 'saved'.

A disadvantage of the system is that if the transducers and wires are exposed they can be damaged by vandals; this might apply in a school building but of course covering and protection of the equipment are possible.

We have no first-hand experience of the system but we know that it has been installed in one government building in order to evaluate the performance. However, this was done only recently so that it will be some time before the performance of the system can be assessed. A particularly important problem is the sensitivity of the system, i.e. the interval of time between what is deemed to be a warning reading and the actual stage of collapse. This interval must be greater than the interval between consecutive readings, unless an automatic system involving a data logger and a computer is used.

Another monitoring system currently in use is that described in the *New Civil Engineer*[106]. This is based on measurements of strain in the soffit or other surface of the beam or slab being monitored. The strain is determined by vibrating wire gauges affixed to the member. The principle of the gauge is that the frequency of the wire varies with the stress in it, just as it does in a violin. Thus an increased deflection of a beam would lead to a higher natural frequency of the vibrating wire gauge attached to the soffit of the beam.

The gauges are calibrated by an applied load of known magnitude, i.e. by what one might call a mini-load test. The loads are placed in various predetermined positions, strains are measured, and the modulus of elasticity of the concrete and the continuity effects are found by the usual structural analysis. The relation between the load applied and the strain recorded should remain constant; any variation would give cause for an immediate investigation. It is only right to admit that variations in temperature and in the atmospheric moisture content affect the readings but under controlled conditions in the interior of a building the effects would be small; taking readings at the same time of the day would be of help. Clearly, the applied load at the time of taking the readings must always be the same; this is easy to achieve in a classroom but less so in a warehouse or a store.

The cost per gauge, including installation, is about £30, and one gauge per beam is adequate; only a proportion of the beams is monitored. In addition, a strain recorder, costing some £500, is necessary but this can serve a number of installations.

The system is used in several buildings. In one, six storeys high, there are 84 gauges[106]. The system is reported to work well but, as with the permanent pulse velocity system, we do not know its sensitivity. The difficulties of safety from vandalism are also similar. On the whole, the vibrating wire gauge system looks promising but we have to await its long-term performance.

There exist other systems of monitoring changes in the concrete in service. One of these measures acoustic emissions, i.e. the development of noise associated with the formation of cracks. The principle is old[107] but the technique is only now being perfected so as to pick up the noise level associated with the formation and growth of cracks at the critical stage and not merely in service. There is also the problem of converting the continuously received input into an integrated value which would give warning of impending failure.

Other monitoring systems check the deflection of beams: a large increase in deflection would indicate a change in behaviour which should be immediately investigated.

Adequate experience with any one of these systems is not yet available and it will be a long time before they can be evaluated. An important point is that, since their objective is to monitor the behaviour of the structure, their installation and operation should be done under the supervision of a structural engineer. He is the only person to interpret the output of a monitoring system in terms of the stability and integrity of the structure.

13.2 Permanent strengthening

The previous section dealt with the situation which followed an assessment of a structure whose state was found to be satisfactory for the present but, as is always the case with HAC concrete, uncertain in the long run. There is, of course, possibly another outcome of the initial assessment, and that is that parts of the structure need strengthening or are in a condition such that local failure may occur. We are not dealing here with the case when the structure, or a part of the structure, is patently unsafe and has to be removed and replaced.

Strengthening as a solution to doubt and uncertainty in the interpretation of test results has been used in a number of structures. Indeed, some engineers feel that strengthening of the HAC concrete parts of a structure is the best solution, especially in the case of socially vulnerable buildings like schools, assembly halls or swimming pools. In this manner, the building owner, as well as the engineer who advises him, knows that his building is safe.

Just as every structure is designed individually, so the strengthening system has to be devised to suit the structure, its use, and the existing services. Thus no general guidelines can be given but a brief description of some of the solutions actually used may be of interest. We propose therefore to mention several actual cases of strengthening so far put into practice.

Steel umbrella

One of these is a building at the Stamford House remand home in Hammersmith[108] . The engineer's decision to install permanent strengthening comes really as no surprise when one learns that the beams are of the same type and from the same manufacturer as those at the Stepney school (see Chapter 7) and that they also support a roof over a swimming pool. The strengthening consists of 600mm deep universal beams running parallel to the existing precast units and spanning 12.4 m. The beams rest on existing load bearing brick piers and carry smaller rolled steel joists which support the HAC concrete beams and the topping. We have thus what one might term a steel umbrella which supports the existing roof, and, if any of the HAC concrete beams fail , the topping will remain over the steelwork.

We should note at this stage how important it is for the strengthening joists to be well tied to the end supports of the HAC concrete beams. In this manner, if a HAC concrete beam fails in shear it will still be supported. The whole question of the shear strength of prestressed HAC concrete beams is of importance as a loss of strength on conversion can lead to a loss of bond and hence to shear failure near the ends of the beam.

Beams and props

Care must be exercised in inserting props as, if they push up a beam where it normally sags, a reversed moment may be induced and severe cracking in the top of the beam can result. Props must of course transfer the load all the way down to the foundations without causing overstressing en route.

Strengthening by beams and props within the building is expensive, particularly as it is important to avoid disruption of electrical and mechanical services in the interior of the building.

Structural infill

In addition to strengthening by external supports, there is also a possibility of putting a structural infill between the HAC concrete beams. Details are not available for publication but we may report a proposal[101] to remove the infill pots between HAC concrete beams and then to use the beams as formwork for placing reinforced concrete beams made with Portland cement. In this manner the HAC concrete beams assume the role of pots in the new structure. Such an operation is clearly difficult and costly.

A variation on this theme has been used in Germany[110]. Here, the pots are left in place and are filled with a (pumped) rich mortar or concrete; a reinforcing bar is also inserted. This new reinforced concrete provides the tension zone of what can be viewed as a slab of mixed new Portland cement concrete and old HAC concrete elements. There are many practical difficulties in piercing external walls and cross-beams and in supporting the structure in the interval between the time of placing the new concrete and the time when it is strong enough to support itself. A high level of workmanship is necessary. The additional weight of the new concrete clearly has to be carried by the structure and this may mean further work. Advantages of the system are that finishes, services, and headroom are not affected.

Slab

Another scheme is to replace an existing screed by a structural (reinforced) slab. As mentioned before, it is not possible to generalize about remedial measures as for each building the structural engineer must evolve the best, and possibly the cheapest, solution.

The weight of the strengthening must be kept to a minimum in order to reduce the effect upon the foundations[122]. The installation must be capable of quick and economic erection with a minimum disturbance to services and finishes. In some cases, fire protection may be difficult and the appropriate regulations may have to be relaxed. The same applies to minimum ceiling heights[122].

As we have already said, the strengthening has to be tailored to each individual building but the use of partitions as supporting members is well worth considering; this is mentioned again in Chapter 14. Cost is the criterion, and of course in some cases it may be cheaper to remove some HAC concrete members and to replace them rather than to provide strengthening[122].

The fail-safe system

One more approach to making a suspect structure safe should be mentioned. This is the installation of a fail-safe system. The purpose of such a system is fundamentally different from strengthening as the system is not intended to support all the HAC concrete beams. Steel joists are inserted below the HAC concrete beams with a clearance of some 50 mm to 75 mm. On top of the joists there is placed a wire mesh or plywood so that if one or two HAC concrete units fail they will be supported, pending full repair work. Thus partial collapse can take place without danger to life. The inherent assumption is that all beams will not fail simultaneously, and this is reasonable.

We have stressed the point of view that the decision about the fate of any one building cannot be made by simple inference from the tests performed. As the guidelines[104] of the Institution of Structural Engineers say, no more than engineering judgement is possible.

14. The Present Situation of High-Alumina Cement Concrete Buildings

14.1 Summary of school buildings

We have seen by now the nature of the 'HAC problem' but we do not know yet how big and how serious it is. There is no central record of buildings giving detailed information on the structural materials used. Local education authorities keep plans and specifications of their buildings but these are not intended for rapid consultation in an emergency situation. In any event, it is by no means certain that such specifications would include particulars of the type of cement used by the manufacturers of precast components. Thus we are not yet aware of all the buildings in which there is HAC concrete. But even when this information is in hand, it will take time and skilled manpower to assess and test the strength of all the buildings containing HAC, let alone to take the necessary remedial action.

Thus it will be many months before we know the full magnitude of the 'HAC problem', and at the present stage we can give information only on what has been found so far.

School buildings were obviously an area of immediate public concern, and data on these have been collected by the Department of Education and Science. They refer to England only and the latest available figures are those of 31st January, 1975. Data for Wales, Scotland and Northern Ireland are available from other government departments. Some of these are given later in this chapter but, in all probability, the situation in those countries, on a per capita basis, is somewhat less severe than in England.

The number of schools in England in which HAC concrete has been identified is 1203. The term school, in the present context, refers to schools of all levels, to colleges of further education, teacher training colleges, and to special schools, maintained by local authorities, but the great majority of those affected by the 'HAC problem' were primary and secondary schools; direct grant schools and private schools are not included. It must be realized that in some of these 1203 schools, the amount of HAC concrete is small, in others there may be several buildings involved. Of this number, 533 have been classified, after a preliminary assessment, as being outside the high-risk category of the Department of the Environment; the classification was given in Chapter 9. Most of these buildings have only short-span HAC concrete beams and may be looked at again later.

Buildings in a further 425 schools have been declared safe after testing or after remedial work has been done. Unfortunately, figures to distinguish between 'safe now' or 'safe after remedial work' are not available. However, it is only reasonable to point out that much of the remedial work may have been very minor. All the buildings in this category are subject to periodic examination.

In 163 schools there are buildings which are in use but which are shored up, either pending further tests or pending remedial work. In a further 145 schools there are buildings which are closed either completely or in part pending further tests. And finally, in 37 schools there are buildings which are closed, wholly or in the affected part, pending remedial work. This is usually major and may include the removal and replacement of HAC concrete beams.

In addition, there are 533 buildings which may contain HAC but have not yet been investigated. To see the proper significance of this number we should add that one-half of these are in one specific part of the country where the local authority lacks the staff for the investigations. While appraisals throughout the country proceed with all possible speed, the 'HAC problem' will be with us, in all likelihood, for another couple of years, and the monitoring and possible further remedial work for many years longer.

The situation in England is then that, of 1203 schools known to contain HAC or of 1736 schools which do or may contain it, 182 have buildings closed in whole or in part. Of these, 37 definitely need remedial work.

To see this number in the right perspective we should mention that the total number of major school projects built in England by local authorities since the Second World War up to 30th June, 1974 was 20,820. Some of these are entirely new schools, others are major extensions. Thus, HAC is involved in about one in twenty of the post-war schools; this ratio takes no account of the many older schools. Moreover, in a large number of schools which contain HAC, only a part of the school is affected so that the proportion of school children who have been disturbed by the 'HAC problem' is very much smaller than one in twenty. We would hazard a guess, but this is only a guess, that no more than one school child in two hundred has been disturbed as a result of the presence of HAC in his or her school.

14.2 Local authorities

In addition to schools, there is a large number of other buildings in which HAC has been used: blocks of flats, old people's homes, private dwellings, hospitals, office blocks. Because many, if not most, of these are not operated by the local authorities it is difficult to obtain figures. However, according to the Department of the Environment it appears that in the United Kingdom as a whole there are between 25,000 and 50,000 buildings (some 20 million square metres) which contain some HAC, and which therefore should be looked at. We cannot give more detailed information on their status but what we can do is to describe the situation in several selected local authorities. We may add that they often differ in their approach and in the solution adopted. The information below refers to all classes of buildings controlled by the local authorities.

Birmingham

In **Birmingham,** up to the end of October 1974, there were about 20 educational buildings closed as a result of the 'HAC problem'. The DTA tests (see Section 12.2) carried out on suspect beams showed in most cases conversion ranging between 80 and 95 per cent. However, there was often variation even within a single beam. Nevertheless, it appears that at least one-half of all the tests have yielded a degree of conversion between 90 and 95 per cent. In general, a higher degree of conversion was found in HAC concrete which was near hot water pipes or even near electric switches. Ultrasonic pulse velocity readings were taken but they were found to be very variable so that not much reliance could be placed on them. Likewise, although a number of cores has been cut, the test results on them have not been found useful, mainly because the cores were very small, often between 25 mm and 50 mm in diameter. We should note, however, that some of the cores registered strengths as low as 14 N/mm^2. Inspection of the cut out concrete often showed a hard top layer but a very soft matrix inside: much of it could be removed with the fingers. We should add that the HAC concrete affected did not all come from the same manufacturer.

When the degree of conversion was established by the DTA, the concrete was classified, using the chart of the Department of the Environment (see Fig. 9.1) for the appropriate age of the concrete. In most cases, the concrete fell into the suspect category and a beam was removed for testing in the laboratory.

In the test, the beam was subjected to cycles of loading and unloading, with a progressively increasing maximum load, and the deflection was measured. In many cases, the load-deflection curve became non-linear at or just above the design working load. This limit of proportionality virtually coincided with the point at which the deflection no longer returns to the extent laid down in the Code of Practice CP 110:1972[86]. In some cases, the beams actually showed distress at about the service design load, the range being between 0.9 and 1.4 of the service load. It is necessary to point out that the definition of distress is somewhat subjective. We should add that the above are results of tests on individual beams; in addition, load tests in situ have been carried out in some cases. While information from all the various types of tests is considered of use, the determination of the actual strength of floors with pots and screeds is thought to be particularly valuable.

In addition to school buildings, some old people's homes have been found to be affected by the 'HAC problem'. There are also some council houses in which HAC has been used but in these, so far, no action has been taken. It is estimated that the extent of the 'HAC problem' in private buildings is similar to that in the public sector.

Decisions on the action to follow the tests are notoriously difficult. In general, in Birmingham, when the degree of conversion was high and the strength was low, some remedial action was deemed necessary but this clearly depends on the details of the building concerned.

Warwickshire

In Warwickshire, up to the middle of October 1974, twelve schools were affected by the 'HAC problem'. Of these, one has been closed completely and one partially. It should be added that in the school which was closed, in addition to weaknesses arising from the use of HAC, there are some serious problems in the foundations of one part of the school.

DTA tests are carried out on all suspect beams and the ultrasonic pulse velocity through the concrete is measured. When the degree of conversion is judged to be high, load tests are applied. This, of course, is not easy, especially when an entire roof or floor is tested.

Some typical values for age, conversion, and ultrasonic pulse velocity, respectively, may be of interest:

School A: 8 to 12 years; not less than 60 per cent, usually above 85 per cent; average of 4.16 km/s.
School B: 10 years; 60 to 85 per cent; average of 4.10 km/s.
School C: 9 years; 85 per cent; average of 4.00 km/s.
School D: 10 years; 85 per cent or more; average of a small number of readings: 3.93 km/s.
School E: 17 years; 90 to 100 per cent; average of 3.90 km/s.

Plate 14.1

Typical isolated castellated beams supporting woodwool slabs. A place where the core was drilled can be seen.

In the last-named school, the roof is to be removed and replaced by open-lattice steel joists. The floor is to be supported by a steel grillage.

The majority of the buildings affected have precast beams with infill pots but some of the schools contain isolated castellated beams (see Plate 14.1). These beams are large so that 75 mm diameter cores can be cut, and an attempt is being made to correlate core strengths and ultrasonic pulse velocity readings. In the meantime, beams over 5.5 m in span which are not designed to act in a composite manner are propped.

We mentioned one school in which a building has been closed completely. Here the roof is of the same type as in the Stepney school (see Chapter 7), including the presence of woodwool type panels. A thorough inspection of the roof beams revealed the presence of horizontal hair cracks in many places. Subsequently, the cracks were found to be as much as 25 mm deep in places. One of the beams had a longitudinal crack at the junction of the top flange and the web, running for over one-half of the beam length (see Plate 14.2).

A surprising discovery was made when the thickness of the flanges was measured: this ranged between 38 mm and 75 mm, the nominal thickness

Plate 14.2 **A crack discovered in an HAC concrete beam during an investigation.**

Plate 14.3 **An example of variable flange thickness and poor compaction together in one beam.**

being 51 mm (see Plate 14.3). The variation must have been caused by misplacing of the side blocks in the moulds as the sum of the thicknesses of the top and bottom flanges always remained constant at 102 mm. This situation makes one wonder about the quality control of precast beams and reminds us that the risk of failure may be aggravated by the presence of HAC but may well be due to other causes also. What we mean, and we have made this point before, is that a badly made beam may be weak but adequate; it is only when the concrete in it deteriorates that the risk of failure becomes high. And we should note that Portland cement concrete does not deteriorate with age.

Plate 14.4

A floor beam prior to load testing. The beam has been exposed to eliminate the effects of composite action.

Plate 14.5

One method of applying a uniformly distributed load for a load test in situ – dustbins full of water.

Another unexpected finding was that in many of the beams there were areas of extremely poor compaction; one of these is illustrated in Plate 14.3. Presumably the quality control was not that expected in a good factory product.

After the inspection and some tests, the roof of the school was subjected to a load test. Excessive deflections were observed and, as a result of this and of other tests, the roof was ordered to be removed. Plate 14.4 shows how a floor had to be removed in order to expose beams for a load test. The arrangements for the load test on a floor in another building are shown in Plate 14.5.

Plate 14.6

An example of poor alignment between the HAC castellated main beam and a secondary beam.

Plate 14.7

Hot water pipes passing through an HAC concrete beam. The white stains on the beam are from the ordinary Portland cement screed above.

There are several points of interest arising from the Warwickshire investigation of HAC concrete members. We mentioned the castellated beams (shown in Plate 14.7); some of them supported heating pipes which passed through the openings. One would hope that the effects of both the heat and the weight of the pipes have been taken into account in the design.

In a number of buildings, there was a mixture of HAC beams and *similar* beams made with Portland cement. Such a practice makes it even more difficult to know which buildings contain HAC. As HAC concrete is different from Portland cement concrete in many respects, we believe that a mixed use is not desirable. We shall return to this topic when describing a structure in Section 14.3.

Bad quality control has already been mentioned. There has also been , in some cases, poor workmanship in aligning the secondary beams with the bearings on the main beams; one such case is illustrated in Plate 14.6. Another example of poor workmanship is shown in Plate 14.7, where Portland cement grout can be seen to have leaked onto the HAC concrete from the screed above.

The cost of the work carried out in Warwickshire and of supplying temporary accommodation up to the middle of October 1974, as a result of the HAC investigations, is estimated to be £185,000 but much still remains to be done.

Plate 14.8 **Faulty bearing of an HAC concrete beam.**

Sheffield

The situation in Sheffield is much better. The DTA test has been used extensively to determine the degree of conversion. Samples are removed by three methods: using a percussion drill, using a normal rotary drill, and by chipping a piece of the beam and grinding the sample of the paste with a pestle and mortar. It appears that significant differences in the results of the DTA test were found between the various methods. This subject was considered in Section 12.2.

Ultrasonic pulse velocity measurements were also taken but they were not conclusive and the test has been discontinued.

No schools have been closed in Sheffield but, in some, HAC concrete beams have been propped. One roof showed an excessive deflection and has been ordered to be removed.

An X7 beam, 4.75m long, was found to have undergone conversion of 90 per cent at the age of $9\frac{1}{2}$ years. The beam was removed and tested to destruction; the load factor was found to be 2.1.

Some HAC concrete beams which were to be used in a structure still under construction in July 1974 (when the use of HAC was discontinued) are to be tested in similar manner, although their degree of conversion is expected to be lower. The load factor at this early age will be of interest.

Newham

In the London Borough of **Newham,** one school has been partially closed. The building, three storeys high, was found (by the DTA test) to contain HAC concrete converted between 50 and 95 per cent; the building is six years old. In consequence, two beams were removed and tested to destruction, one on site and one in the laboratory. The ultimate strength of both beams was found to be satisfactory and the building has been put back into service with a permanent monitoring system (see Section 12.7). An example of a faulty bearing discovered during the investigation is shown in Plate 14.8.

In the second school, the degree of conversion was found to range between 20 and 50 per cent, and tests on three individual beams showed a satisfactory ultimate strength. The building is four years old. It has been put back into use with a permanent monitoring system.

In Newham, the ultrasonic pulse velocity method is not thought to be of value. Cores are not used either. Instead, attempts have been made to determine the compressive strength of concrete on parts of the beams tested, using an approach similar to the equivalent cube test of the British Standard BS 1881:Part 4:1970. There are complications arising from the shape of the beam section and from the presence of reinforcement, so that the technique still needs perfecting.

In addition to schools, five other buildings (flats and offices) are under investigation in Newham but no results are yet available.

Croydon

Croydon is prominent in its dealing with the 'HAC problem' because on 26th February, 1974, the Borough Council imposed a ban on the further use of HAC for structural work both in public and in private building projects. This was forward-looking action. It resulted not only from the failures at Camden and at Stepney but also from the fact that the 'HAC troubles' in Croydon started in 1973, when a small number of schools and blocks of flats were identified as containing structural members made with HAC.

In one school, built in 1954, the DTA test showed the degree of conversion to be between 40 and 75 per cent. The HAC beams in the building were mainly of rectangular cross-section and spanned 7.9 m. The beams had been under surveillance for about 18 months when, in July 1973, the deflection of some of them was found to have increased considerably, in some cases up to 100 mm. It was decided to close the classrooms, and subsequently the roof was removed and reconstructed with steel joists. The Engineering Division of the Croydon Borough Council worked in close consultation with the Building Research Establishment, and several of the beams removed were sent there for tests. One of these failed at the design working load.

Following the circular [92] of the Department of the Environment of 20th July, 1974, the buildings containing non-composite HAC construction were provided with intermediate supports to the beams. The programme of regular surveillance, including DTA and deflection measurements, which was instituted in 1973, has been continued.

Lancashire

The authorities in **Lancashire** have kindly provided very full data on the HAC situation there. It is estimated that possibly 5000 buildings contain HAC. The figures of interest, at the beginning of December 1974, are as follows.

In educational buildings:-
62 have been identified as containing HAC
17 have beams with spans under 5 m; no action is taken on these
29 have beams with spans over 5 m
16 are of composite construction (any span)
22 have been temporarily propped
5 have been declared safe
2 are waiting to be propped.
In other buildings:-
240 are thought possibly to contain HAC
25 have been identified as containing HAC
17 have beams with spans under 5 m
10 have beams with spans over 5 m
8 are of composite construction (any span)
1 has been temporarily propped
9 are waiting to be propped.

Isolated beams are tested as follows. The DTA test and ultrasonic pulse velocity determination are carried out, and on the basis of these the beam with the highest degree of conversion is selected for the load test in situ. The load factor used is 1.8 on the assumption of fully converted strength as given by the Code of Practice CP116:Part 2:1969. Specially constructed trollevs carrying a load of 1.5 tonnes are used. The procedure is similar in the case of beams acting compositely but, in addition, cores are taken from the screed in order to determine the ultimate load of the composite section as built.

From Lancashire, let us move to a nearby BOROUGH, where there are the following buildings known to contain HAC: 20 schools, 20 meeting halls and old people's homes or clubs, 80 factories or offices, 3 bridges, and 20 houses.

Ultrasonic pulse velocity measurements have not been taken but the DTA test is used in all cases. This has shown that, typically, the degree of conversion in buildings more than seven years old is between 70 and 90 per cent; in buildings three to four years old, the degree of conversion is around 45 per cent. The DTA test is followed by a load test. In the majority of cases, this has yielded satisfactory results.

There are a few 'special cases'. Three buildings have isolated beams similar to those in the Stepney school (see Chapter 7). They have been closed down and full-scale load tests are being carried out. In one school the design is similar to that of the Camden school (see Section 6.1), and it has been decided to increase the length of the bearing of the beams. Two buildings have been found to have very bad roof leaks; props have been installed pending further investigations.

Liverpool

It is estimated that in the **Liverpool** area there are about 2000 buildings containing HAC. Up to now, 45 schools have been identified as containing the material. Of these, ten use isolated beam construction and are being subjected to load tests. The remaining 35 schools are continuing in service, at least for the time being.

The situation in two schools is of interest. One has a roof of a type similar to that in the Camden school (see Section 6.1), with beams spanning 6.4 m and resting on bearings 25mm long. It is thought that these bearings have to be increased.

In the other school, which is ten years old, DTA tests have shown the degree of conversion to be approximately 90 per cent. Main roof beams, spanning 16.8m, were subjected to a load test and have been found to be satisfactory. However, the purlins have caused concern. They are of T-section, with 127mm by 51mm flange and overall height of 153mm. Under a load test on site, minimal deflection was observed but some cracking was also observed, and it is likely that remedial action will be taken.

The load test used in Liverpool is more severe than that prescribed by the Department of the Environment: the total load applied is 1.5 times the dead plus live load, instead of 1.25 times the same value (see Section 12.5). Bricks are used to apply the load.

In a school where HAC concrete beams are used compositely, the DTA test showed the degree of conversion to be about 90 per cent but the majority of load tests gave satisfactory results. However, in one area, the 305mm by 127mm rectangular roof beams showed deflections up to 150mm. It is proposed to strengthen this roof by means of steel joists.

Trying to give a more general picture of the Liverpool area, we can say that the vast majority of the HAC concrete beams there came from one manufacturer. They all show conversion, often of between 80 and 90 per cent in large beams, somewhat less in small ones. The attitude of the local authority is to investigate each school building rather than to prop all HAC concrete beams as a matter of principle. This is an understandable approach because there is in the area a very large number of dwellings containing HAC, in which the spans are under 5m. Like all authorities Liverpool is awaiting guidance from the Building Research Establishment.

In the Liverpool area there is a large number of Catholic schools. These are being appraised by outside consultants. Many of the schools were closed until cleared by a load test.

Staffordshire

In **Staffordshire,** over 300 buildings have been inspected to see whether they contain HAC. Of these, 12 have been found (up to the end of November 1974) to contain HAC, and were investigated fully.

The main tool in the first instance was DTA. The degree of conversion established in each case is related to the graph of the Department of the Environment (see Fig. 9.1), bearing in mind the age of the concrete. Detailed visual examination was also made, particular attention being paid to the deflection of the prestressed concrete units and a regular programme of check levelling on beam soffits was introduced. The ultrasonic pulse velocity method and cores are not used.

The investigations showed that most of the HAC concrete falls into the *"suspect"* category of Fig. 9.1, and some is *"highly converted"*.

On the basis of the tests, certain HAC concrete units were selected for load tests, which are proceeding. Other units were ordered to be permanently propped or to be removed, with temporary props being immediately installed.

One of the buildings seriously affected was an assembly hall, about 12m times 24m in plan. There was inadequate headroom to insert supports for the HAC concrete roof, which was removed with a mobile crane. Open-lattice proprietary joists were then placed on the old bearings and a roof consisting of timber bearers and woodwool units was installed. The job took three weeks and cost approximately £5,000, inclusive of the replacement of the ceiling. This case illustrates the point that a desire to strengthen a roof or a floor, rather than to replace it, may be vitiated by inadequate headroom.

One minor structure that had to be removed was an elevated long-span canopy where prestressed HAC units were found to sag as well as to curl anticlastically, i.e. in the direction transverse to the span. The canopy was consequently dismantled.

Five buildings, four of them multi-storey, are to be permanently supported. Details of the supports vary from building to building but, as a rule, open-lattice joists are placed below the HAC concrete units. These joists are bolted to the existing columns. Spanning between the joists, there are placed steel channel or angle sections, which in turn support a wire mesh.

The system was originally to serve as a fail-safe device, with stresses approaching the elastic limit, for use where floors had camber and where it might be expected, in the light of all the relevant factors, that no future overall deterioration would occur. This type of grid system is not designed to take the full weight of the HAC concrete; the purpose of the grid is to support a single HAC concrete beam, or two , which may fail and to restrain the debris from falling on the floor (or the people) below.

However, it was decided to design the supporting grid so that it will carry the full dead and superimposed load; the increase in cost necessary to achieve this is relatively small. The wire mesh will be in two layers: the upper one a lightweight electro-galvanized 12mm square mesh to retain any debris, and the lower layer a heavier 75mm by 225mm galvanized mesh, which suitably restrained at the ends, would carry the full load of a failed beam. With this system, both the wire meshes and the supporting steel sections would become permanent supports if any floor deteriorated in the future to such an extent that it would need replacement. The latter would then be relatively simple.

It is proposed to insert an asbestos-based material between the floors and the supports. Such a pack will consolidate slightly under the pressure of the deflecting floor and thereafter translate the load directly to the steelwork, and will also provide fire resistance.

Sefton

The magnitude of the task in the Metropolitan District of **Sefton** (Merseyside) is similar to that in Staffordshire. Over 200 school buildings were inspected; of these, 34 were found to contain HAC. After further tests, one building was completely closed, one was permanently strengthened, and four were temporarily propped.

A considerable number and variety of tests have been carried out, including DTA, ultrasonic pulse velocity measurements, core tests, and load tests. However, because it was felt that it was not possible to predict reliably the future behaviour of existing HAC concrete members, it was decided to use the tests only for the purpose of determining the priority to be given for permanent strengthening works. These will be carried out in a phased programme.

Several school buildings merit description. In one high school, which was of typical prestressed concrete beam and infill pot construction, the DTA showed in the majority of cases a degree of conversion of 75 to 80 per cent, with some readings down to 50 per cent, at the age of 5 years. Cracking in both the ceiling and the floor was observed, indicating excessive deflection.

Plate 14.9 **One method of permanent propping in use in a school.**

There was no hesitation in deciding to strengthen the building permanently.

Steel joists were inserted below, and parallel to, the HAC concrete units and connected to the existing steel columns by brackets, which were welded to them. On top of the steel joists and at right angles to them, there were placed timber joists. To ensure bearing of the HAC concrete units on the timber joists, timber wedges were driven (see Plate 14.9). Such an operation requires care as a forced excessive upward movement of the HAC concrete units may damage them seriously.

A primary school in Sefton was of rather different construction. Here the main hall roof was supported on isolated prestressed HAC concrete beams 0.6m to 0.9m deep, spanning 12.2m to 13.7m; the beams are shown in Plate 14.10. The DTA test showed 70 to 80 per cent conversion, at the age of 16 years, and the values of the ultrasonic pulse velocity were low. It was decided to strengthen the roof and, as a temporary measure, timber supports were placed (see Plate 14.10).

Plate 14.10 **Temporary timber propping in use in a school hall.**

In the classroom building of the same school, there were prestressed concrete floor beams of rectangular cross-section. The DTA and ultrasonic pulse velocity test results were similar to those for the roof of the main hall. Further inspection followed. When the false ceiling was removed it was noticed that parts of the bottom flanges of the HAC concrete units had spalled, the debris actually lying on the false ceiling. In consequence, some of the prestressing steel was exposed and was corroding. Temporary timber supports were therefore inserted in all the classrooms and cloakrooms, as shown in Plate 14.11. In the light of experience it will be decided how long they will remain in position or whether another supporting system should be installed.

One of the schools in Sefton may be mentioned by name since it has been extensively described in the press. This is the Hatton Hill school, where there is a large number and a large variety of HAC concrete beams. The roof of the main hall is supported by prestressed concrete beams, rectangular in cross-section, 457mm by 229mm, and spanning approximately 12.2m. The roofs of the classrooms are supported by the more conventional I-beams in some cases, and by rectangular beams in others.

The DTA tests indicated a high degree of conversion in all cases, about 75 to 80 per cent at the age of 14 years. In many parts of the classroom area, there were obvious signs of water leakage through the roof and the deflection of some of the beams was excessive. Concrete from these beams was found (by the DTA test) to have undergone sulphate attack. Further tests followed: ultrasonic pulse velocity determinations, cores, and strength test of two beams removed from the structure. The test results were not easy to interpret

Plate 14.11 **Temporary timber propping in use in the cloakrooms of a school.**

but the minimum cube strength for the concrete in the whole structure was estimated to be 34 N/mm^2. In consequence, it was thought that there was no immediate danger of collapse, providing future chemical attack does not induce a significant loss of strength. Thus, it was decided to replace flat roof areas, where severe chemical attack is known to have taken place, and to strengthen the pitched roofs of the assembly hall and of the classrooms, where leakage has not been significant due to better drainage conditions. The strengthening was done by steel joists and channels. Plate 14.12 shows the plaster to the HAC concrete beams being removed prior to the installation of the supporting joists. The absence of plaster will remove a cause of chemical attack but we may wonder why gypsum plaster was put in contact with HAC concrete in the first instance.

Plate 14.12 **A plaster screed being removed from an HAC concrete beam in a school hall prior to propping.**

Two more schools in Sefton may be mentioned. In one, there was a main hall and a gymnasium,both similar in construction to the gymnasium at the Stepney school (see Chapter 7). The hall and the gymnasium have been closed pending permanent strengthening.

In the other school, 11 years old, the DTA showed a degree of conversion of 55 to 85 per cent in prestressed HAC concrete open-lattice girders; many of these are isolated. On the basis of ultrasonic pulse velocity measurements, it was estimated that the minimum cube strength is 21 N/mm^2 and it was concluded that there was no danger of collapse, providing no localised chemical attack occurs. However, shortly afterwards a fire broke out in the school. Subsequent tests showed a significant drop in the ultrasonic pulse velocity (from between 4.40 km/s and 4.55 km/s to approximately 3.5 km/s). The future of the building is still uncertain.

And last, but of course not least, we mention **Leeds.** Here over 500 buildings have been inspected, 40 of them in detail, i.e. they were subjected to some tests. Twelve school buildings have been identified as containing HAC; five of these belong to the local authority and seven are Catholic schools. All these buildings have been closed during testing, temporary propping being considered uneconomical.

As far as non-school buildings are concerned, it is estimated that there are at least 100 buildings containing HAC; five of these are in one shopping complex.

The general testing procedure in Leeds is to survey the deflections and to follow this by the DTA test on about one beam in ten. Likewise, ultrasonic pulse velocity measurements are taken on about one beam in ten. If possible, cores are cut to correlate with the ultrasonic pulse velocity measurements. As appropriate, some beams are removed and tested to destruction in the laboratory. No load tests on site have been carried out so far.

Detailed information on two schools may be of interest. In one, built in 1964, the DTA test showed a very high degree of conversion in the majority of isolated roof beams, spanning 5.5m. Consequently, laboratory tests on some of the beams with low ultrasonic pulse velocity values have been carried out.

The loading sequence was as follows. First, a bending moment of 13.0 kNm was applied for 24 hours. After removal of load, the deflection was found to recover 80 per cent on the average. The beam was then tested to destruction. The values of the ultimate moment for four beams ranged from 13.7 kNm to 15.8 kNm, the design working moment, according to the manufacturers, being 14.5 kNm. The calculated ultimate moment, on the basis of the design data, but allowing for a slight reduction in the strength of concrete, was 31.0 kNm. In consequence, it was decided to remove all the beams and to replace them with steel joists.

In another school, built in 1959, isolated roof beams have been found to have a high degree of conversion. Some beams were removed and tested in a manner similar to that described in the preceding paragraph. The ultimate strength of the beams was found to be adequate. However, the bearing arrangement of the beams was similar to that at the Camden school (see Section 6.1) and the length of the bearing will be increased by appropriate remedial measures.

The total cost of repairs of these two schools is likely to reach £450,000.

In a third school, there is a gymnasium and an assembly hall, each similar in design to the buildings at the Stepney school (see Chapter 7). Isolated HAC concrete beams spanning 12.2m were used. Deflection was found to vary between 35mm and 90mm under dead load only. In consequence, it was decided to replace the roof beams. Other beams in the school have yielded poor ultrasonic pulse velocity measurements and appear to show a high degree of conversion. The school has been closed and testing is continuing.

Many other schools, built between 1959 and 1965, show a high degree of conversion.

The shopping complex, referred to earlier, is six years old. The roof members show a low degree of conversion. On the other hand, the HAC concrete in the beam and pot infill floor is highly converted, probably because underfloor heating has been used. It is hard to understand how HAC concrete could have been employed with this form of heating.

14.3 Other buildings

Information on buildings not controlled by local authorities is difficult to obtain, but to give a typical picture of the situation we can report on some of the buildings which fall within the purview of the Property Services Agency, Eastern Region. This region contains rather less than one-seventh of government buildings in the United Kingdom. The regional office is checking about 2000 buildings. Of approximately 300 buildings examined, 15 have been found to contain HAC concrete. Proposals for strengthening two have been agreed and the others are still being considered.

The decision to strengthen was taken, in the majority of cases, on the basis of the DTA tests. The Agency engineers view the ultrasonic pulse velocity test as of some value in locating the weaker members in a building, but not in the assessment of the strength of concrete. Likewise, it is thought that cores can help in obtaining a general picture of the state of a building.

Since the buildings of the Property Services Agency are different in character from school buildings, it is of interest to describe them in some detail.

Two of the buildings are large halls housing expensive equipment. They were constructed in 1959-1960 and have roofs consisting of isolated precast prestressed rectangular HAC concrete beams supporting woodwool slabs. The beams span 10m, and the two halls are 110m and 70m long, respectively. Shortly after the construction was completed, one of the roof beams collapsed owing to the failure of the end nib which rested on a bearing. On examination it was found that the top non-prestressed shear reinforcement stopped short of the nib, which was thus made of plain concrete. Supporting steel brackets were then installed, but 6mm below the beam soffit so that they came into action only after the nib failure. In a period of 14 years, 40 per cent of the nibs have failed in one of the halls and 13 per cent in the other.

Plate 14.13 **Beam on the supporting bracket after failure of seating.**

Originally, the roof beams were grouted into their seating on the main beams made with Portland cement. However, the main portal frame beams showed some sideways displacement, the effect being that the HAC concrete beams pulled away from their seatings; the resulting gap can be seen in Plate 14.13. The HAC concrete beams also lost a considerable proportion of their strength: cores showed that the present strength ranges between 14 N/mm^2 and 27 N/mm^2. In consequence of all this, it has been decided to strengthen the two buildings by placing an open-lattice joist under each HAC concrete beam. The cost for both buildings is estimated to be about £35,000.

Another building for which the Property Services Agency is responsible is a workshop in which a crane rail is supported on nibs in reinforced concrete columns, cast in situ. With astonishment, it has been found that the part of some of the columns in the vicinity of the nib is made of HAC concrete while the remainder of the column, above and below, is made with Portland cement. Moreover, in one column, part of the nib is made with one cement and part with the other. Thus fresh HAC concrete appears to have been placed against fresh Portland cement concrete; for the dangers involved in such an operation we should look at Section 2.4. The DTA test of the HAC concrete has shown a degree of conversion of 80 to 85 per cent. The use of the crane has been suspended pending a full structural appraisal to be followed by load tests.

In situ concrete was also used in the construction of a water tower in 1948. The shell roof of the tower has a thickness of 50mm to 100mm. The HAC concrete in the shell has been found to be converted to an average of 85 per cent; it looks dark brown and has a flaky surface. Since there is no danger to life, replacement or strengthening has been given a low priority.

An unusual structure is a hangar, some 60m in span, whose roof is supported by prestressed precast concrete segmental arch ribs. These consist of alternating large segments made of Portland cement concrete and of smaller segments of HAC concrete. Cores have shown strengths of the HAC concrete between 6.2 N/mm^2 and 8.3 N/mm^2. The HAC concrete is soft and it can be removed by a water jet or by sandblasting; it is proposed to use one of these methods and to replace the HAC segments with Portland cement concrete. The cost of the scaffolding necessary to support the structure during this operation is estimated at £12,000 and the total cost of repairs at £60,000.

The Eastern Region of the Property Services Agency controls also two office blocks which are of interest to us. In one, involving HAC concrete beams and infill pots, the degree of conversion at the age of ten to twelve years has been found to range between 70 and 85 per cent. No other tests have been carried out so far but there is visual evidence of chemical attack where leakage through woodwool slabs has occurred, and temporary props have been installed. In the other office block, the appearance of the beams was better but the degree of conversion was found to be even higher. On one of the floors, a heavy load had been applied by the occupants, possibly in excess of the design load. Examination of the beams indicated that during construction some damage to top flanges occurred, as shown in Plate 14.14.

Some of the information on HAC concrete buildings which are being investigated, and often strengthened, comes from consulting engineers who advise clients on the work necessary. Many of these consultants are firms who did not use HAC in their designs in the first instance. They deal with

Plate 14.14 **Damage to the top flange of an HAC concrete beam supporting an excessive load from above.**

schools, office blocks, factories, workshops, old people's homes, and private houses. A few of the cases may be worth mentioning here.

In a bank floor, where HAC concrete I-beams span 4.0m to 4.9m, the DTA showed a degree of conversion of 45 to 55 per cent at the age of one year. Visual inspection showed the beams to be acceptable, and the building continues to be occupied but is to be inspected visually at six-monthly intervals. However, the consulting engineer is considering possible corrective measures.

In commercial premises, the floors are made of I-section HAC concrete beams with a 64mm composite screed; the beams span 10.7m. At the age of six years, the degree of conversion was found to range between 55 and 85 per cent. Visual inspection suggests that the floors are in a reasonable condition, and it is likely that they will be allowed to continue in service.

The roof of the same premises consists of I-beams, which sag typically 20mm in a span of 11.0m. The DTA has shown the degree of conversion to be only between 25 and 50 per cent and, paradoxically, this is of some concern: one can expect further conversion and further loss of strength, so that periodic checks will be necessary.

In a swimming pool building, whose roof was supported by isolated HAC concrete beams spanning 12.2m, the beams were removed and replaced by concrete beams made with Portland cement. Tests showed extensive conversion. In another swimming pool building, where HAC concrete beams were spanning 5.3m and were supported on a steel structure spanning over the pool, the DTA showed a degree of conversion of 85 per cent. It was decided to remove four of the beams for tests in a laboratory; the outcome of these is not yet known.

In some two-storey houses, there were discovered isolated HAC concrete beams, spanning 4.5m, and supporting the floors and the roof. In the course of maintenance work, it was thought wise to prop the HAC concrete beams.

A building with a different use, and therefore possibly of interest, is a students' hall of residence in London. The building is eight storeys high. The floors are of the common HAC concrete beam and infill pot type, spanning 7m. At the age of 15 years, the degree of conversion was found to be 80 to 90 per cent. One of the floor beams was taken out and tested in the laboratory. The ultimate strength of the beam corresponded to a factor of safety of 1.38 on the design service load. Compression tests on sawn-off flanges of the broken beam indicated that, at least in places, the top flange was weaker than the bottom. This has also been found the case in other beams, made by different manufacturers.

Deflection readings taken throughout the building at quarter-span suggest that all the beams are similar in quality. It appears that the concrete topping, although not designed to act compositely, contributes to the structural action of the beams so that, in fact, there is some composite action. The HAC concrete beams are supported by reinforced concrete walls but there exist at mid-span non-load-bearing partitions, which support a high proportion of the load. One possible way of strengthening the building is to make these partitions load bearing; the building would be monitored thereafter. The alternative of substituting cast-in-situ reinforced concrete for the infill is also being considered.

A use of HAC concrete which does not fit into the general pattern is a chocolate factory in the Midlands. The case attracted attention because the HAC concrete beams in an advanced stage of deterioration were shown in the *This Week* programme on ITV in October 1974.

The beams in question were reinforced and arranged as a grid to support warm water pipes forming part of a cooling system. The HAC concrete

Plate 14.15 **HAC concrete beams at the chocolate factory, showing extreme deterioration.**

beams were placed over a pond in the open and were supported by Portland cement concrete piles; these have remained sound. The warm water pipes had perforations through which water was discharged forming a spray. This allowed cooling, and the water from the pond was re-circulated. The system was in use only when the weather was not cold, i.e. six to eight months in the year. The advent of more modern cooling methods led to the system being abandoned after several years; since then the beams have not been in use and have carried only their self weight.

In the middle of 1974, in the wake of interest in HAC, the beams were inspected. The colour of the HAC concrete was chocolate brown (appropriately for a chocolate factory) and it was found to have deteriorated to such an extent that it could be broken away by hand (see Plate 14.15).

It is obvious that HAC is not an appropriate material for use in this type of service but, as pointed out in Chapter 5, this knowledge was not as widespread as it should have been. Not much is known about the quality of the concrete used: it was site mixed in winter and placed in a small-job manner under very difficult conditions over water.

14.4 Scotland and Northern Ireland

To show that 'HAC problems' occur all over the United Kingdom we propose to mention two schools in Scotland; two otherswere referred to in Chapter 9. One of these [116] was a school in Dundee, from whose roof nine HAC concrete beams and two Portland cement concrete beams were removed and tested in the laboratory. The beams spanned between 6.25m and 6.86m, were of T-cross-section, and were nominally 254mm deep. They were not designed to act in a composite manner. The beams were ten to twelve years old.

The appearance of the beams was poor, and they were sagging in places to such an extent that there was extensive ponding of water on the roof. This was discovered, however, only during the investigation arising from the search for HAC concrete.

In a test to destruction, all the HAC concrete beams failed by crushing of the concrete in the compression zone. The range of the calculated average compressive strength of the concrete at failure was 13.0 N/mm^2 to 19.6 N/mm^2. Assuming this stress to be 0.6 of the cube strength, these figures yield estimated cube strengths of 21.6 N/mm^2 to 32.7 N/mm^2. The loss of prestress since transfer was estimated to be 40 to 60 per cent. The Portland cement concrete beams were estimated to have lost 40 to 50per cent of their prestress but their equivalent cube strength was of the order of 40 N/mm^2. Thus they were much further from failure in compression, or indeed from failure, than the HAC concrete beams. All the beams were removed and are being replaced.

In the other school, also in Dundee, prestressed HAC concrete purlins, 125mm in depth, spanned approximately 4m. Three of the beams were removed and tested in the laboratory, where their strength was found to be similar to that of beams from the first school. We should note, however, that the span was much shorter. The decision on the future of the roof is not known.

In Northern Ireland, the Works Division of the Department of Finance identified about 90 buildings containing precast prestressed HAC units. The use of HAC in in situ work seems to have been very small. The majority of the buildings with HAC have shallow (152mm) I-beams with infill pots, designed either for composite action with structural topping or for non-composite use. The buildings vary in type and use, ranging from factories to small blocks of flats. None of them is more than three years old, the majority are under two years.

Because of the small size of the HAC concrete units, cores could not be cut. Some DTA tests were performed but, in view of the young age of the concrete, they were even less informative than usual. Some laboratory tests on the strength of individual beams were made but no general conclusions have been reached. In the meantime, temporary props have been inserted in a number of buildings.

A small number of buildings contain isolated I-beams, 305mm deep, spaced at about 3.6m and forming part of a precast building frame system. The buildings vary in age up to ten years and include several schools. The DTA tests have indicated a high degree of conversion. One building in a school has been closed pending further investigations and in some of the remaining schools temporary props have been installed. As in England, tests continue. Northern Ireland has been spared nothing.

14.5 The Overall Picture

The overall picture of the 'HAC problem' in the whole of the United Kingdom is being assembled by the Building Research Establishment. Their data are confidential until the final report has been published some time in 1975. But certain general features have emerged. Most of the concrete referred to the Establishment, and all that is more than seven years old, has been found to have been converted to a very high degree. Even a few months' old beams showed a degree of conversion of 70 per cent. The strength of the concrete tested ranged between 10 N/mm^2 and 50 N/mm^2, but the characteristic strength in many cases was no more than 20 N/mm^2 to 25 N/mm^2. Of the buildings of various types investigated up to December 1974, nearly one hundred are in some sort of distress; roofs are more often affected than floors. Many of these are of course the same buildings as those referred to earlier under the heading of various local authorities.

Epilogue

From the discussion of the various tests it is apparent that, in the vast majority of cases, the engineer is unable to examine and test HAC concrete to a point where he can guarantee safety. He can express a professional opinion but this is not enough for the owner or for the local authority to accept responsibility for the integrity of a building. Thus there is no one willing to put his name to a health-certificate. Consequently, in virtually all cases of buildings containing HAC, some remedial measures will be undertaken. This view has been expressed in a note[132] to the Concrete Society, and is probably a realistic assessment of the situation. We may note that in Bavaria exactly the same attitude was adopted. It is likely that short-span beam and pot infill construction is safe but, without guidance, the engineer assessing such a building is in an exceedingly difficult situation.

The main value of the various tests is thus to determine the order of priority of the work to be done. With time, we shall see more props, fail-safe devices, and certainly more inspections of existing structures. The need to inspect buildings, and especially roofs, is an important lesson learnt from the 'HAC trouble'.

But there are other lessons. One is that manufacturers' claims about their products must not be uncritically accepted. The British Standards Institution has served the country well but the balance of manufacturers, users, and general interest parties (government agencies and researchers) has not been always well maintained. Even if on paper the balance is right, in practice the attendance by those more directly interested may be such as to sway the vote. The American practice of qualifying the votes by the interest category may be worth emulating.

Another lesson is that the designer should at all times be in full control of his structure. This applies to the detailed properties of the materials used, as well as to workmanship, so that the move toward performance specifications, just now in vogue, may be unwise.

Even more unwise is construction of the type where a building is 'delivered' to fulfill a certain function without a structural engineer being consulted about the materials, connections, tolerances etc. This package deal seems economically attractive but is much less so when all aspects of safety are taken into consideration.

The panic that followed the HAC collapses is to be regretted. Clearly, buildings cannot be *guaranteed* to be safe if they are to be constructed economically, and possibly even if they are not. We do not expect our cars or railways to be 100 per cent safe. We do not construct our sea defences or river levees to protect us from the highest imaginable flood. But we want our houses to be as safe as houses (although the origin of the expression refers to our trading institutions, alas no longer safe). This may be so because we associate risk with dynamic situations and want absolute safety when sitting at home. Such safety is not practicable, and we must persuade the public that structural design is based on an acceptable element of risk.

But the risk, to be acceptable, must be very small and must not be compromised by a knowing acceptance of unstable materials or of uncertain manufacturing techniques. There was ample evidence that HAC may lead to trouble in our buildings, there was the example of so many other countries which did not use HAC in structures. And it is not even as if HAC were an indigenous product made of British raw materials or by a British-owned manufacturer.

A great many engineers refused to use HAC in their designs and many architects used precast concrete systems not involving HAC concrete units. But others ignored all warnings. May I quote from the conclusions to my paper published in 1963; these were as fair to HAC as possible to avoid a charge of anti-HAC bias. *"HAC concrete with a water-cement ratio higher than about 0.5 will lose a considerable proportion of its strength under ordinary conditions existing outdoors in Europe, including England, over a period of 20 to 30 years. Many structures are expected to have a life in excess of this period.*

The loss of strength is accelerated by a rise in temperature and humidity, and the cement should not be used in buildings where a high humidity and considerable warmth, say in excess of 25 °C, are expected. Such conditions may sometimes not be anticipated at the time of construction but may occur at a later date, e.g. when an industrial building is put to a different use from that for which originally designed . . . the new user may not even be aware of the fact that HAC cement had been used in construction. Even short periods of higher temperature, especially coupled with a rise in humidity, are detrimental because their effect is cumulative and irreparable. This danger of loss of strength at some time in the future is an argument against the use of HAC cement."

And, after a discussion on the loss of strength in various mixes, we pointed out that, even with a water-cement ratio of 0.35, conversion *"may lead to an inadequate load factor in prestressed concrete, and the use of high-alumina cement in this type of construction is not recommended".*

And the final words were: *"HAC is an essentially different binder from Portland cement, so that a comparison of concretes on the basis of strength above may be very misleading."*

With all this before us, we should ask: why did we use HAC in our structures?

Perhaps someone will answer this question. In the meantime, we must beware of transferring our experience with HAC concrete to concrete in general. Concrete made with Portland cement has served us well and remains a sound, durable and economic structural material. HAC has its place, too, but this is in special applications and not in structures.

References

1 T.W. PARKER. The constitution of aluminous cement. *3rd. Int. Symp. Chem. Cement,* London, 1952, pp. 485-515.

2 F.M. LEA. Chemistry of Cement and Concrete. Revised edition of the "Chemistry of Cement and Concrete" by F.M. Lea and E.H. Desch. Arnold, London, 1956, revised edition 1970, 727pp.

3 T.D. ROBSON. High-alumina cements and concretes. Contractors Record Ltd., London,1962, 263 pp.

4 H. LAFUMA. Quelques aspects de la physico-chimie des ciments alumineux. *Rev. Gen. Sci. Appl.,* Vol. 1, No. 3, 1952, pp. 3-11.

5 F.M. LEA. Cement and concrete. Lecture delivered before the Royal Institute of Chemistry, London, 19th December 1944.

6 T.D. ROBSON. High alumina cement. *Arch. Sci. Rev.,* Vol. 11, No. 1, March 1959, pp. 49-56.

7 W.H. GLANVILLE and G.F. THOMAS. Further investigations on the creep or flow of concrete under load. *Build. Res. Tech. Paper, No. 21,* London,1939, 44pp.

8 A.M. NEVILLE and H.W. KENNINGTON. Creep of aluminous cement concrete. *4th Int. Symp. Chem. Cement,* Washington D.C., 1960, pp. 703-708.

9 A.M. NEVILLE. Properties of Concrete. Pitman, London,1963, revised edition 1972, 686 pp.

10 BLUE CIRCLE GROUP, CEMENT MARKETING CO. LTD. Lightning concrete for strength and chemical resistance. Technical note 72 3, December 1972, 11pp. (reprint of technical note 71 3, December 1971).

11 N.G. ZOLDNERS and V.M. MALHOTRA. Discussion on reference 45. *Proc. Instn. Civil Engrs.,* Vol. 28, May 1964, pp. 72-73.

12 LAFARGE ALUMINOUS CEMENT CO. LTD. Ciment Fondu: concreting in cold weather. Technical data sheet no. SA2, October 1973, 3pp.

13 BRITISH STANDARDS INSTITUTION. The structural use of precast concrete. Code of Practice CP 116:Part 1 (Imperial units): 1965, 156pp., CP 116:Part 2(Metric):1969, 156 pp.

14 H.G. MIDGLEY. The mineralogy of set high-alumina cement. *Transactions, British Ceramic Society,* Vol. 66, No. 4, 1967, pp. 161-187.

15 R. KELLY. Solid-liquid reactions amongst the calcium aluminates and sulpho-aluminates. *Canad. J. Chem.,* Vol. 38, 1960, pp. 1218-1226.

16 R. ALÉGRE. Etude des effects sur les ciments alumineux hydratés de la transformation de $CaO.Al_2O_3.10H_2O$ sous l'action de la température. *Revue des Materiaux de Construction,* Vol. 630, 1968, pp. 101-108

17 A.M. NEVILLE. Further tests on the strength of high-alumina cement concrete under hot wet conditions. *RILEM Int. Symp. on Concrete and Reinforced Concrete in Hot Climates,* Haifa, 1960, 22pp.

18 V. MATIC. Uticaj starosti na betone sa aluminatnim cementom. *Saopstenja Instituta za Ispitivanje Materijala,* Belgrade, Vol. 8, September 1960, pp. 22-32.

19 W. ALBRECHT and H. SCHMIDT. Einpressmörtel für Spannbeton. *Dtsch. Auss. Stahlbeton,* No. 142, 1960, 44pp.

20 R. JONES and E.N. GATFIELD. Testing concrete by an ultrasonic pulse technique. *Road Research Tech. Paper 34,* H.M.S.O., London, 1955, 48pp.

21 S.J. SCHNEIDER. Effect of heat-treatment on the constitution and mechanical properties of some hydrated aluminous cements. *J. Amer. Ceram. Soc.,* Vol. 42, No. 4, April 1959, pp. 184-193.

22 P. LHOPITALLIER. Calcium aluminates and high-alumina cement. *4th Int. Symp. Chem. Cement,* Washington D.C., 1960, pp. 1007-1033.

23 L.S. WELLS and E.T. CARLSON. Hydration of aluminous cements and its relation to the phase equilibria in the system lime-alumina-water. *J. Res. Nat. Bur. Stand.,* Vol. 57, No. 6, December 1956, pp. 335-353.

24 F.M. LEA. Effect of temperature on high-alumina cement. *Trans. Soc. Chem. Ind.,* Vol. 59, 1940, pp. 18-21.

25 J. TALABÉR. Durabilité des ciments alumineux. *RILEM Int. Symp. on the Durability of Concrete,* Prague 1961, Final report 1962, pp. 109-114.

26 H. LAFUMA. Conditions de la prise, du durcissement et de la conservation des ciments alumineux. *Rev. Mat. Construction Trav. Publ.,* No. 305, February 1935, pp. 29-30.

27 BUILDING RESEARCH STATION. High-alumina cement. Digest No. 27, February 1951, 6pp.

28 N. DAVEY. Influence of temperature upon the strength development of concrete. *Build. Res. Tech. Paper,* No. 14, H.M.S.O., London, 1933, 76pp.

29 G.C. HAGGER. The use of aluminous cement in the construction of the Mosul tunnel, Iraqi State Railways. *J. Instn. Civil Engrs.,* Vol. 25, 1945-1946, pp. 142-149.

30 F.M. LEA and N. DAVEY. Deterioration of concrete in structures. *J. Instn. Civil Engrs.*, Vol. 32, 1948-49, pp. 248-275.

31 P.H. BATES. Some properties of high-alumina cement from six countries. *First commun. new Int. Ass. for Test. Mat.*, Group B, Zurich, 1930, p. 211.

32 A.M. NEVILLE and I.E. ZEKARIA. Effect on concrete strength of drying during fixing electrical resistance strain gauges. *RILEM Bulletin*, No. 38, 1957, pp. 95-96.

33 A.M. NEVILLE. Effect of warm storage conditions on the strength of concrete made with high-alumina cement. *Proc. Instn. Civil Engrs.*, Vol. 10, June 1958, pp. 185-192.

34 K. NEWMAN. Design of concrete mixes with high-alumina cement. *Reinf. Conc. Rev.*, Vol. 5, No. 5, March 1960, pp. 269-294.

35 S.C.C. BATE. Report on the failure of roof beams at Sir John Cass's Foundation and Red Coat Church of England Secondary School, Stepney. *BRE CP 58*, June 1974, 18pp.

36 O.J. MASTERMAN. High-alumina cement concrete with data concerning conversion. *Civil Engng. and Public Works Rev.*, Vol. 56, No. 657, April 1961, pp. 483-486.

37 F.G. THOMAS. Influence of time upon strength of concrete. General report on first session of RILEM Symposium on "Influence of time upon strength and deformation of concrete". *RILEM Bulletin*, No. 9, December 1960, pp. 17-34.

38 F.M. LEA and C.H. WATKINS. The durability of reinforced concrete in sea water. 20th report of the Sea Action Committee of the Instn. of Civ. Eng. D.S.I.R., *Nat. Build. Stud. Res. Paper 30*, H.M.S.O., London, 1960, 42pp.

39 A. HUMMEL et al. Versuche über das Kriechen unberwehrten Betons. *Dtsch. Auss. Stahlbeton*, No. 146, 1962, 133pp.

40 D.G. MILLER and P.W. MANSON. Long-time tests of concretes and mortars exposed to sulphate waters. *Technical Bull. 194*, University of Minnesota, Agricultural Experiment Station, 1951, 111pp.

41 K. OKADA. Unpublished report on tests at Onoda Cement Co., Japan.

42 T.D. ROBSON. Discussion on reference 34. *Reinf. Conc. Rev.*, Vol. 5, No. 5, March 1960, pp. 294-295.

43 ANON. No more wetness about warmth? *New Civil Engineer*, 21st March 1974, pp. 40-43.

44 M. DURIEZ. Traité de Matériaux de Construction. Éditions du Moniteur des Travaux Publics, Paris, 1957.

45 A.M. NEVILLE. A study of deterioration of structural concrete made with high-alumina cement. *Proc. Instn. Civil Engrs.*, Vol. 25, Paper 6652, July 1963, pp. 287-324.

46 A.M. NEVILLE. Closure to the discussion on reference 45. *Proc. Instn. Civil Engrs.*, Vol. 28, May 1964, pp. 78-84.

47 ANON. Alkaline hydrolysis the likely culprit. *New Civil Engineer,* 21st February 1974, p. 10.

48 E. FREYSSINET and A. COYNE. Observation sur une maladie des bétons de ciment fondu. *Genie Civil,* Vol. 90, No. 11, 1927, pp. 266-269.

49 R. L'HERMITE. Personal communication.

50 R. CAVENEL. Reparation du Pont de la Corde, sur la Penzé près des Carantec (Finistère). *Ann. Ponts. Chauss.,* 1944 (Paper written 1941).

51 CHEF DE SERVICE, Centre Expérimental de Recherches et d'Études du Batiment et des Travaux Publics, Paris. Communication of 27th March 1956.

52 F.L. HARWOOD. Discussion on reference 29. *J. Instn. Civil Engrs.,* Vol. 26, 1945-46, pp. 613-614.

53 F.M. LEA. Some studies on the performance of concrete structures in sulphate-bearing environments. "Performance of concrete", edited by E.G. Swenson, University of Toronto Press, 1968, pp. 56-65.

54 P. HAKANSON and H. LINDQUIST. Erfarenheter av aluminacement. Appendix 1 to Kommentarer till 1960 Års Cementbestammelser. *Statens Betongkommitté,* KB 1, Stockholm, 1961, pp. 19-23.

55 A.C. HARTLEY. Discussion on reference 30. *J. Instn. Civil Engrs.,* Vol. 32, 1948-49, pp. 276-277.

56 B.F. SAURIN. Personal communication.

57 R.B. KIRWAN. Discussion on reference 30. *J. Instn. Civil Engrs.,* Vol. 32, 1948-49, pp. 573-574.

58 C.K. HASWELL. Discussion on reference 30. *J. Instn. Civil Engrs.,* Vol. 32, 1948-49, pp. 282-284.

59 A.E. BUTLAND. Personal communication.

60 G. REHM. Schaden an Spannbetonbauteilen aus Tonerdeschmelzzement. *Zement-Kalk-Gips,* Vol. 17, No. 3, March 1964, pp. 102-111.

61 A.E. SHEYKIN and F.M. RABINOVICH. Strength of high alumina cement and factors influcncing. (In Russian). *Doklady Akademii Nauk* USSR, Vol. 177, No. 6, 1967, pp. 1407-1410.

62 A.M. NEVILLE. Tests on the strength of high alumina cement concrete. *Journal of New Zealand Engineering,* Vol. 14, No. 3, March 1959, pp. 73-76.

63 H.H. MÜLLER. Personal communication.

64 J.P. MITCHELL. Discussion on reference 45. *Proc. Instn. Civil Engrs.,* Vol. 28, May 1964, p. 60.

65 G.H. SADRAN. Discussion on reference 45. *Proc. Instn. Civil Engrs.,* Vol. 28, May 1964, pp. 76-78.

66 J. BOBROWSKI. Discussion on reference 45. *Proc. Instn. Civil Engrs.,* Vol. 28, May 1964, pp. 62-65

67 ANON. Collapse due to suspect concrete says BRE report. *Surveyor,* 5th July 1974, p.4.

68 H. LAFUMA. Calcium aluminates in aluminous and Portland cements. Lecture delivered at the Institute Eduardo Torroja in Madrid, 30th May 1963.

69 H. RÜSCH. Personal communication.

70 R.E. DAVIS. Personal communication.

71 P.J. FRENCH, R.G.J. MONTGOMERY and T.D. ROBSON. High strength concrete within the hour. *Concrete,* Vol. 5, No. 8, August 1971, pp. 253-258.

72 DEPARTMENT OF EDUCATION AND SCIENCE (prepared by BRE). Report on the collapse of the roof of the assembly hall of the Camden School for Girls. H.M.S.O., London, 1973, 16pp.

73 A.M. NEVILLE. Creep of concrete: plain, reinforced and prestressed. North-Holland, Amsterdam, 1970, 622pp.

74 H.G. MIDGLEY and K. PETTIFER. Electron optical study of hydrated high-alumina cement pastes. *Transactions, British Ceramic Society,* Vol. 71, No. 3, 1972, pp. 55-59.

75 F.J. SAMUELY AND PARTNERS. Preliminary comments on Messrs. Ove Arup and Partners report on the investigations into the partial roof collapse in the Geography room, University of Leicester, Bennett Building, May 1974, 3pp.

76 INSTITUTION OF STRUCTURAL ENGINEERS. Report on the use of high-alumina cement in structural engineering. August 1964, 17pp;

77 MINISTERE DES TRAVAUX PUBLICS, FRANCE. Circulaire Série B, No. 5, 27th January 1928.

78 MINISTÈRE DES TRAVAUX PUBLICS, FRANCE. Circulaire Série B, No. 27, 25th March 1935.

79 MINISTÈRE DE LA PRODUCTION INDUSTRIELLE ET DES COMMUNICATIONS. Secretariat Général des Travaux et des Transports. Circulaire Série A, No. 1, 5th January 1943.

80 MINISTÈRE DE LA CONSTRUCTION, FRANCE. Circulaire No. 70-31, 5th March 1970.

81 MINISTERS FÜR LANDESPLANUNG, WOHNUNGSBAU UND OFFENTLICHE ARBEITEN NORDRHEIN–WESTFALEN. Verwendung von Tonerdeschmelzzement. Publication IIB2 - 2.323, No. 179/62, Ministerialblatt NRW, issue A 15, No. 95, 31st July 1962, p. 1413.

82 A. MARKESTAD. Personal communication.

83 T. RECHARDT. Personal communication.

84 C.S. FORUM. Personal communication.

85 COMITÉ EUROPÉEN DU BÉTON. International Recommendations for the design and construction of concrete structures. June 1970, 80pp.

86 BRITISH STANDARDS INSTITUTION. The structural use of concrete. Code of Practice CP 110, November 1972, 154pp.

87 DEPARTMENT OF THE ENVIRONMENT. Building and Buildings, Second Amendment to the Building Regulations, Proposed BRA/938/3, 9th August, 1974, Statutory Instrument No. 1944, made 22nd November, 1974, laid before Parliament 9th December, 1974, put into operation 31st January, 1975, 44pp.

88 DEPARTMENT OF THE ENVIRONMENT. Collapse at Camden School for Girls. Circular letter of 27th June 1973, 1p.

89 DEPARTMENT OF THE ENVIRONMENT. Roof collapse at Camden School for Girls. Circular 109/73, 17th August 1973, 8pp.

90 DEPARTMENT OF THE ENVIRONMENT. Collapse of roof beams - The Sir John Cass's Foundation and Red Coat Church of England Secondary School - Stepney. BRA/1068/2, 28th February 1974, 6pp.

91 DEPARTMENT OF THE ENVIRONMENT. Collapse of roof beams - The Sir John Cass's Foundation and Red Coat Church of England Secondary School - Stepney. BRA/1068/2, 30th May 1974, 4pp.

92 DEPARTMENT OF THE ENVIRONMENT. Collapse of roof beams - The Sir John Cass's Foundation and Red Coat Church of England Secondary School - Stepney. BRA/1068/2, 20th July 1974, 12pp.

93 SCOTTISH DEVELOPMENT DEPARTMENT. High alumina cement concrete in buildings. *SDD Circular,* No. 40, BC/BSR/4 /6/12, 2nd August 1974, 9pp.

94 PROPERTY SERVICES AGENCY, Department of the Environment. High alumina cement concrete. *Technical Instruction,* Serial CE 100 B 62, August 1974, 10pp.

95 LAFARGE ALUMINOUS CEMENT CO. LTD. Some of the properties and applications of Ciment Fondu. 1961, 16pp.

96 LAFARGE ALUMINOUS CEMENT CO. LTD. Ciment Fondu: mix proportioning. *Technical data sheet* S2, July 1973, 2pp.

97 D.E. SHIRLEY. Principles and practice in the use of high alumina cement. *Municipal Engineering,* Vol. 145, No. 4, 19th January 1968, pp. 120-124 No. 5, 26th January 1968, pp. 143-146.

98 C.D. CROSTHWAITE. Discussion on reference 45. *Proc. Instn. Civil Engrs.,* Vol. 28, May 1964, pp. 71-72.

99 BRITISH STANDARDS INSTITUTION. Structural use of prestressed concrete in buildings. Code of Practice CP 115:Part 1 (Imperial units):1959, 44pp, CP 115:Part 2 (Metric): 1969, 44pp.

100 PROPERTY SERVICES AGENCY, Department of the Environment. High alumina cement. *Technical Instruction* Serial CE 100 B 62, 4th December 1974, 14pp.

101 D.D. DOUBLE and A. HELLAWELL. Unpredictable strength of HAC. *Building,* 6th September 1974, p. 117.

102 BUILDING RESEARCH ESTABLISHMENT, Department of the Environment. BRE statement on high alumina cement concrete testing methods. Press release, October 1974, 2pp.

103 M.H. ROBERTS and S.A.M.T. JAFFREY. Rapid chemical test for the detection of high-alumina cement concrete. *BRE Information sheet* IS 15/74, September 1974, 2pp.

104 INSTITUTION OF STRUCTURAL ENGINEERS. Guidelines for the appraisal of structural components in high alumina cement concrete. HAC/1/1974, October 1974, 17pp.

105 ANON. Putting HAC to the test. *New Civil Engineer,* 26th September 1974, p. 20.

106 A.S. SAFIER. A practical approach to HAC concrete appraisal. *New Civil Engineer,* 10th October 1974, p. 32.

107 H. RÜSCH. Physical problems in the testing of concrete. Cement and Concrete Association Translation No. 86 of *Zement-Kalk-Gips*, Vol. 12, No. 1, 1959, pp. 1-9.

108 P. REINA. Raising steel umbrella under prestressed beams. *New Civil Engineer,* 9th May 1974, p. 37.

109 ANON. Repairs on Stepney school completed. *Building*, 27th September 1974, p. 47.

110 G. REHM and D. BRIESMANN. Moeglichkeit zur Wiederherstellung der Tragfaehigkeit beschaedigter Fertigteiltraegerdecken. *Beton,* Vol. 18, No. 7, July 1968, pp. 255-258.

111 A.M. NEVILLE. Discussion on reference 34. *Reinf. Conc. Rev*, Vol. 5, No. 7, September 1960, pp. 461-462.

112 B. MAYFIELD and M. BETTISON. Ultrasonic pulse testing of high alumina cement concrete: in the laboratory. *Concrete,* Vol. 8, No. 9, September 1974, pp. 36-38.

113 R. ELVERY. HAC tests should allow for microcracking and stress history. *New Civil Engineer,* 5th September 1974, pp. 19-20.

114 J.H. BUNGEY. Ultrasonic pulse testing of high alumina cement concrete: on the site. *Concrete,* Vol. 8, No. 9, September 1974, pp. 39-41.

115 B. DUXBURY. UPV concrete testing - a blind date. *New Civil Engineer,* 8th August 1974, p. 32.

116 A.R. CUSENS and N. JACKSON. Kirkton High School, Tests of prestressed concrete beams. Report prepared for Ove Arup and Partners, Dundee. Civil Engineering Department, University of Dundee, May 1974, 48pp.

117 E. KUPZOG, K.J. LEERS and E. RAUSCHENFELS. Der pH-Wert von hydratisierten Calciumaluminaten und Tonerdezement. *Tonindustri-Zeitung,* Vol. 90, No. 4, 1966, pp. 155-161.

118 H.G. SMOLCZYK. Die röntgenographische Beurteilung von Beton aus Tonerdezement. *Betonstein-Zeitung,* Vol. 30, No. 11, 1964, pp. 573-579.

119 F.K. NAUMANN and A. BÄUMEL. Bruchschäden an Spanndrähten durch Wasserstoffaufnahme in Tonerdezementbeton. *Archiv. für das Eisenhüttenwesen,* Vol. 32, No. 2, 1961, pp. 89-94.

120 BUILDING RESEARCH ESTABLISHMENT. Annual Report 1973, H.M.S.O., London 1973, pp. 15-16.

121 E. SKIBSTRUP. Personal communication.

122 J.E.C. FAREBROTHER. Note to the Concrete Society, 3rd December 1974.

123 ANON. Lancs. to safety check 5000 buildings. *Building,* 2nd August 1974, p. 27.

124 AGRÉMENT BOARD. Certificate No. 69/32 of 14th April 1969 for Swiftcrete ultra high early strength Portland Cement, 11pp.

125 P.K. MEHTA and G. LESNIKOFF. Conversion of $CaO.Al_2.O_3.10H_2O$ to $3CaO.Al_2O_3.6H_2O$. *Journal of the American Ceramic Society,* Vol. 54, No. 4, April 1971, pp. 210-212.

126 O.J. MASTERMAN. Discussion on reference 45. *Proc. Instn. Civil Engrs.,* Vol. 28, May 1964, pp. 57-60.

127 J.T. KAY. Discussion on reference 45. *Proc. Instn. Civil Engrs.,* Vol. 28, May 1964, pp. 60-61.

128 DEPARTMENT OF THE ENVIRONMENT. High alumina cement concrete in buildings. BRA/1068/2, 2nd January 1975, 3pp.

129 R. PIÑEIRO. Personal communication.

130 K. OKADA. Personal communication.

131 R. TSUKAYAMA. Effect of conversion on properties of concrete using high-aluminous cement. *Proc. 5th Int. Symp. Chem. Cement.* Tokyo, 1968, Part 3, pp. 316-327.

Index